Sand in Construction

Edited by Sayed Hemeda

Published in London, United Kingdom

Sand in Construction
http://dx.doi.org/10.5772/intechopen.97977
Edited by Sayed Hemeda

Contributors
Rajab Abousnina, Rochstad Lim Allister, Metta Achyutha Kumar Reddy, Veerendrakumar C. Khed, Maria Stefanidou, Parthena Koltsou, Fadhel Aloulou, Habib Sammouda, Serhan Haner, Murat Aydın

First published in London, United Kingdom, 2022 by IntechOpen
IntechOpen is the global imprint of INTECHOPEN LIMITED, registered in England and Wales, registration number: 11086078, 5 Princes Gate Court, London, SW7 2QJ, United Kingdom

British Library Cataloguing-in-Publication Data
A catalogue record for this book is available from the British Library

Additional hard and PDF copies can be obtained from orders@intechopen.com

Sand in Construction
Edited by Sayed Hemeda
p. cm.
Print ISBN 978-1-80355-585-0
Online ISBN 978-1-80355-586-7
eBook (PDF) ISBN 978-1-80355-587-4

Meet the editor

Sayed Hemeda received a Ph.D. in Civil Engineering from Aristotle University of Thessaloniki, Greece. Currently, he is the manager of heritage science programs at the Liberal Arts and Culture Center (LACC) and a professor at the Basic and Applied Science Institute (BAS), Egypt-Japan University of Science and Technology (E-JUST). He is the first professor of Heritage Science and Architectural Preservation of Architectural Heritage, Conservation Department, Faculty of Archaeology, Cairo University, Egypt. He is also the former manager of historic buildings, Conservation Center of Archeology, Cairo University. Dr. Hemeda is the recipient of many awards from Cairo University including prizes for scientific excellence (2017), the Scientific Encouragement Prize (2014), and the best Ph.D. thesis (2009–2010). He was also awarded the General Union of Arab Archaeologists prize for academic excellence (2019). He has published 85 articles and 29 international books and has been cited 330 times. He has given more than 58 invited lectures in 16 countries. His primary interests are geotechnical engineering for architectural heritage preservation and engineering data analysis including pattern recognition as applied to primarily analytical data from various sources such as objects of cultural significance. He is editor-in-chief of the International Journal of Advances in Geological and Geotechnical Research. He is an editorial board member for many organizations and publications, including the Open Journal of Geology, Progress of Electrical and Electronic Engineering, and Geoscience Journal. He is also a scientific and organization committee member for many international conferences .

Contents

Preface

Sand is a type of naturally occurring material that is granular, loose, and fragmented, consisting of particulate matter such as rock, coral, shells, and so on. Sand is typically finer than gravel but coarser than silt.

This book discusses some contemporary issues related to sand in construction in engineering projects, which are critical components in civil construction and often require detailed management techniques and unique solutions to address failures and remedial measures.

This book brings together a collection of chapters covering construction problems and solutions. Chapter 1 is concerned with the role of sand in mortar's properties. It discusses the diachronic presence of mortars and emphasizes the role of aggregates to understand their impact on the longevity and durability of mortars.

Chapter 2 discusses the development of bentonite-modified concrete and reviews the literature about the physical and chemical properties of bentonite, bentonite-blended cement mortar, bentonite-modified cement concrete, and reinforced concrete. It also presents the history and development of bentonite-modified concrete.

Chapter 3 examines the effects of mill conditions on breakage parameters of quartz sand in the district of Şile on the Black Sea Coast of İstanbul. The authors obtained the specific rate of breakage values from the particle size distributions acquired after various grinding periods. As a result of grinding tests, an increase in the rate of breakage is observed due to the increase in ball diameter.

Chapter 4 presents the Thermal Conductivity and Mechanical Properties of Organo-clay-Wood Fiber in Cement-Based Mortar. This paper is orientated to study the compressive resistance and thermal conductivity of compressed and stabilized clay blocks in the cement matrix. The effect of the content of wood fiber (WF) became studied as a reinforcement material in cement mortars. The porosity, compressive energy, thermal conductivity, and the composite of cement hydration had been investigated.

Chapter 5 discusses the sources, properties, remediation, and engineering and construction applications of oil-contaminated sand. Oil leakage during the exploration, production, and transportation of crude oil is considered to be a significant issue worldwide because crude oil spills severely impact the environment and also affect the physical and chemical properties of the surrounding soil.

Chapter 6 presents a study of a building project consisting of nine floors, eight flats, and two elevators. The study uses Building Information Modelling (BIM) and Industry Foundation Classes (IFC) and Industry Foundation Classes eXtensible Markup Language (IFCXML) data. BIM is recognized as the most effective platform for information exchange on building projects in the construction industry.

It supports the development of various software. It facilitates automated or semi-automated ACCC of the building projects for compliance with building regulations and standards for the participants (designer, architect, engineer, contractor, owner, etc.) involved in the building production process.

I would like to express my gratitude to IntechOpen for their efforts in publishing this book and to the contributing authors for their work and patience.

Sayed Hemeda
Professor,
Faculty of Archaeology,
Conservation Department,
Cairo University,
Giza, Egypt

Liberal Arts and Culture Centre (LACC),
Egypt Japan University of Science and Technology (E-JUST),
Egypt

Chapter 1

The Role of Sand in Mortar's Properties

Maria Stefanidou and Parthena Koltsou

Abstract

Mortars are diachronic composite materials used in masonry construction to serve multiple roles. Their durability and esthetic harmonization in constructions of the different eras were the reasons why numerous research works have been realized over recent decades. Each time, the role of the mortars' components revealed significant pieces of information on the technology used. Despite the indisputable role of the binders on the mortar's quality, aggregates of different characteristics had a significant role in the behavior of mortars. The addition of aggregates to a binding system in mortars technology has proved to confer technical advantages as they contribute to volume stability, durability, and structural performance. Apart from the different types of aggregates, as their mineralogy and origin are concerned, the volume content in the mixture, the maximum size, and their gradation influences the structure of a binder—aggregate mixture and the performance of mortars overall. In the present article, the diachronic presence of mortars is presented. The role of aggregates is emphasized to understand their impact on the longevity and durability of the mortars.

Keywords: sand, mortars, composites, mechanical-physical properties

1. Introduction

Mortars are among the first building materials used in constructions, even from prehistoric times. Their study reveals a great source of information regarding the evolution of their technological characteristics and application techniques, the availability and exploitation of raw materials, as well as the wider socioeconomic aspects of each era. In any case, it seems that ancient masons were fully aware of the significant role of mortars in constructions and could exploit the raw materials that were available along with the application techniques [1]. In particular, the role of the quality of aggregates on the properties of old mortars has been known since, at least, Roman times. Natural sands of different origins and nature (river, quarry, sea) and crushed bricks combined with binders which were usually lime-based, were used for many centuries (**Figure 1**). These mortars were of different types and served as bedding, renders or plasters, floors, and mosaics' substates forming masterpieces of the world cultural heritage [2].

It is evident from the classic authors that the Romans preferred sharp sands to rounded sands, as they knew that these would produce stronger mortars; for example,

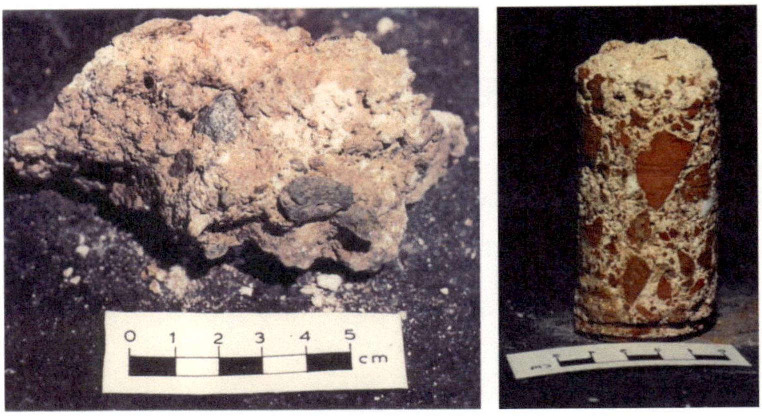

Figure 1.
Coarse aggregates of natural origin in bedding mortars of the fourth century AD (left) and crushed brick as aggregates in a bedding mortar of the sixth century AD (right).

Palladius, Pliny, and Vitruvius refer to recipes and guidelines for criteria that can be used for sand selection in the mixtures [3–5]. Among the requirements they mention, the origin, the shape, and the cleanness of the sand are the ones that prevail. They noticed the direct relation of the sand quality to the setting and strength of the mortars, and they gave precise directions to avoid, for example, sea sand due to salt contamination that can accelerate the weathering of the mortar. Manufacturing mortar was the first milestone in building history which has been continuously evolved up to the modern concrete. The materials used for mortar manufacture since antiquity were binders (clay, lime, pozzolan, brick/tile dust), aggregates (sand, gravel, crushed brick, pumice), and materials that were less frequently found and used (such as chopped straw, egg whites, reeds, blood, palm fibers, milk, and goat hair.). In Akrotiri of Thera, Greece (1700–1400 BC), structural mortars were made of local origin clay, mixed with gravel, charcoal, and straw [6]. In Hellenistic monuments, such as Dilos residences (second century BC), lime-pozzolan mortars were mainly found, with aggregates of natural origin and of granulometry mainly 0–8 mm [7]. During the Roman period (second century BC–third century AD), the use of lime and pozzolan dominated in constructions, while brick dust and crushed brick also started to be used [8]. The systematic and in high proportion use of brick dust and crushed brick in lime or lime-pozzolan mortars were expanded during the Byzantine era fourth–fifteenth century AD [9]. Aggregates (natural origin and crushed brick) were of gradation 0–8 mm to 0–16 mm, with a B/A ratio 1/2–1/3 [10]. The effectiveness of the adhesion between the binder and the crushed brick aggregates achieved in those cases was impressive. During the Ottoman period (fifteenth–nineteenth century AD), structural mortars were manufactured by using the available raw materials [11]. They were mainly lime-based (pure lime or lime with clay), while in specific constructions demanded in resistance to humidity (baths, cisterns), pozzolan and brick dust were also added. Aggregates were of natural origin (in some cases crushed brick was also added), of 0–8 mm granulometry, and of B/A ratio 1/2 [12]. In Medieval times (fifteenth–nineteenth century AD), structural mortars mainly consisted of lime (in some cases pozzolan was added), natural or crushed aggregates, and crushed brick in gradations 0–4 mm to 0–8 mm and B/A ratio 1/1–1/2 [13]. During the nineteenth and beginning of the twentieth century, structural mortars varied depending on the building type and the local constructional tradition. Aggregates were usually of the natural origin of 0–8 mm gradation [14]. Later, scholars such as Lanas [15] referred to the importance of the binder/aggregate interface as a zone that requires special attention. From the historic research of the components of mortars, it is obvious that the presence of sand was catalytic and continuous. In relation to the origin of the sand used, it was mainly local, from streams or rivers, and in special cases, crushed bricks or tiles in different gradation were added [16, 17]. Aggregates are the most ubiquitous materials in construction. Nowadays, the building industry uses aggregates as materials for construction, mainly in their bound state with cement to form concrete, bitumen to form asphalt, or as components for composite materials. Nevertheless, the utilization of aggregates has a long history in construction technology and especially in mortars. Over the last decades, due to the increasing cost of raw materials and the continuous reduction of natural resources, the recycling of industrial waste has become an interesting option for the building industry. Nowadays, many large industries use manufactured sand alone for producing mortar by partially replacing river sand. In these complex systems, the aim seems to be first, the utilization of low-cost materials from local resources and second ensuring the quality and performance of materials for specific applications. Therefore, there is still a continuous usage of sand in construction works. Alternative approaches to completely replacing sand in mortars have intensified over the last decades [18]. At the moment, the increasing awareness of society about safeguarding heritage buildings and at the same time protecting the environment promotes strategies of combining principles of restoration with environmental friendly materials and techniques.

2. The influence of sands in mortar's properties

2.1 Mortars as composite materials

Composites are materials made by combining two or more other materials. These materials are important in the construction sector as building technology has been favored by the advanced properties that composites can offer. The development of composite materials along with related design and manufacturing technologies constitute one of the most important advances in the history of materials. Composites are multifunctional materials having unprecedented mechanical and physical properties that can be tailored to meet the requirements of a particular application [19]. Thus, new achievements have been constructed as the innovative composites could add new possibilities to the engineers' imagination.

Mortars are a specific type of composite material, which consists mainly of three phases—paste as the matrix, interface transition zone (ITZ), and aggregates. The properties of mortars are influenced by:

- Aggregates (type, percentage, shape)
- Binders (activity, percentage)
- Their contact surface area
- Mixing water
- The maintenance conditions of the applied specimens

At fresh state, mortar should be workable (do not break and do not flow), plastic (to have consistency to hold and not flow on overload loads) and it should show volume stability (do not cause contractions or expansions). At harden state, it should have the required strength and the required porosity. Aggregates constitute the strongest phase, hold a significant percentage in the volume of mortar and are frequently used in sand size (up to 4 mm). Conditions for the use of aggregates in mortars is the health of both the parent rock and the grains (without breaks, cracks, impurities), low porosity—small absorption index, homogeneous granulometric grade, percentage of the fines (<0.075 mm) should not exceed 5% [20]. The presence of fines in lime-based mortars can cause considerable alterations to the properties of the mortars. Their presence significantly reduces the strength and increases the volume shrinkage of mortars [21]. Furthermore, the porosity can be increased, and the same can also happen with capillarity when fine aggregates are participating in excess. Additionally, the type of fines also seems to play a role in their behavior in relation to the basic binder. For example, the strength is decreased in compositions with clay fines while porosity is affected mainly by limestone fines [22]. Capillarity also seems to be affected by the type of fines as the compositions containing fines 10–15% presented low absorption probably because fines block capillary pores [23].

In the case of mortars, as composite materials and keeping in mind that aggregates retain the inherent properties of the rocks from which they are derived, it can explain that the color, the chemical and physical characteristics of the aggregates directly affect the specific weight, the measure of elasticity, the volume stability, the appearance, and the mechanical and physical properties of mortars. The addition of sand to a binding system in mortars technology has proved to confer technical advantages as they contribute to volume stability, durability, and structural performance [16]. The gradation of the aggregates was wide, but the most adequate part was sand of 0–4 mm. Coarse aggregates up to 1 cm were used in thick joints [10] and also combined with sand for structural mortars while sand or finer aggregates (0–2 mm) are usually the constituents of the renders or plasters [24]. Usually, aggregates are obtained after the gentle grinding and sieving (based on EN1015-1) [25]. Even homogeneous distribution of grains is usually obtained as shown in **Figure 2** in a typical bedding mortar.

The ratio of binder to aggregates (B/A) ranges widely but it could be said that for most of the structural mortars, it is 1:2.5 or 1:3 while for the renders and plasters are

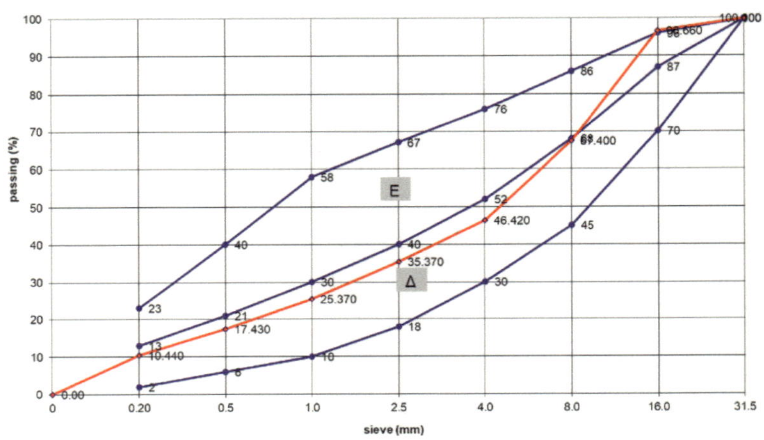

Figure 2.
Typical gradation of old structural mortar.

richer in binder content and the ratio is mostly 1:1 or 1:1.5 [23]. Apart from the different types of aggregates, as their mineralogy and origin are concerned, the volume content in the mixture, the maximum size, and their gradation influences the structure of a binder—aggregate mixture and the performance of mortars overall [10]. The added aggregates strengthen the composite, and the associated interface weakens it. These two opposite effects offset each other, and the combination of them leads to declined strength. Generally, a strong cohesion between the mortar binder and coarse aggregate confirms the good masonry properties. On the other hand, the increase of aggregate content reduces the workability of a mix and thus, reduces the strength as well [26]. It has been mentioned before that aggregate plays a role in restraining the shrinkage of cement paste, and that the shrinkage of the aggregate itself can be neglected [27]. It has been found in various composite materials that a certain amount and proper size of the aggregate are beneficial to the strength and fracture energy of the composite [28]. For mortar specimens, aggregates have a significant influence on both rheological and mechanical properties. Their specific gravity, particle size distribution, shape, and surface texture influence markedly the properties of mortars in the fresh state. On the other hand, the mineralogical composition, toughness, elastic modulus, and degree of alteration of aggregates are generally found to affect the properties of mortars and in the hardened state [29]. The drying shrinkage strains in investigated mortars are changed significantly by different kinds of fine aggregate materials. The water content of the mortar mix proportion is a major factor in drying shrinkage evolution. Increasing the unit water content can result in an increase in the amount of capillary water, and hence more shrinkage strain would be obtained. The bonding stress of the weak interface zone between the coarse aggregate and paste can be improved when a chemical reaction between the aggregates and the paste [30].

More recently, the role of the recycled sand from waste demolition, when examined in mortars, revealed that it was more beneficial in lime mortars rather than in stronger lime-pozzolan or lime-pozzolan and cement mortars as a decrease in their performance were recorded in the latter cases due to the mortars' structure [31]. It seems that two competitive mechanisms acted in these mortars; high porosity (due to high water content and the nature of the aggregates) which assists toward low strength and durability and the chemical reaction due to the presence of reactive components which creates a strong structure. This chemical reaction is a stronger mechanism in the case of lime mortars and prevails in relation to the competitive mechanisms of the higher porosity [32].

In an effort to test different aggregate-related properties to hydraulic lime mortar, Pavia et al. suggested [33] that an increase in the calcite content of the aggregate lowers the flexural and compressive strength of the mortar. At the same time, they proved that sharp aggregate, as well as aggregate with small average particle size, tends to increase the mechanical strength and bulk density of a mortar simultaneously reducing porosity, water absorption, and capillary suction. Additionally, they concluded that aggregates containing particles of a wide size range can increase the mechanical strength and bulk density of the hardened mortar diminishing porosity, water absorption, and capillary suction.

2.2 Different sands in mortars

The role of aggregates on the structure and behavior of lime-based mortars is examined by studying the influence of the aggregate content, type, and grain size on the strength, porosity, and volume stability of the mortars. Trying to understand how

Figure 3.
(a) Black sand, (b) yellow sand, and (c) blonde sand.

the properties of the sands influence important macroscopic properties of pure lime mortar, threes sands that were available in the market were selected and analyzed in the laboratory. All of them were river origins of siliceous content (**Figure 3**).

X-ray diffraction analysis (XRD) using a D2 Phaser 2nd generation, Bruker instruments, indicated that blonde sand was containing quartz, feldspar, magnetite, calcite, hornblende. Yellow sand contained quartz, feldspar, magnetite and black contained quartz, feldspars, biotite, and hornblende. Physical properties, such as water absorption, specific gravity, and sand equivalent (S.E.), are shown in **Table 1** while the chemical analysis revealed the silicic nature of the sands (**Table 2**).

Lime mortars were prepared using lime CL90 (based on EN459) [34] and the compositions were produced, as shown in **Table 3**. The workability was measured with a flow table as described in EN1015-3 [35].

The samples were cured based on EN456 and at 28 days, the compressive strength and the open porosity were recorded (**Table 4**).

The results show that there are different properties recorded in the produced mortars even when siliceous sands are used. The different properties, such as S.E. and the

	Water absorption %	Specific gravity g/cm^3	S.E.
Blonde	0.70	2.36	90.5
Black	1.46	2.35	98.0
Yellow	1.09	2.34	75.0

Table 1.
Physical properties of sands.

	Soluble in acids % b.w.								Soluble salts % b.w.		
Sample	**Na_2O**	**K_2O**	**CaO**	**MgO**	**Fe_2O_3**	**Al_2O_3**	**SiO_2**	**L.I.%**	**Cl^-**	**NO_3^-**	**SO_4^{2-}**
Black sand	3.24	2.63	3.05	1.28	3.66	13.83	68.37	3.45	0.19	0.08	<0.01
Blonde sand	2.97	1.76	3.16	1.87	5.69	14.17	66.00	4.23	0.09	0.11	0.04
Yellow sand	7.82	0.94	0.73	0.46	1.02	18.71	67.14	2.88	0.01	<0.01	0.31

Table 2.
Chemical analysis of sands.

Composition	Lime	Blond sand	Black sand	Yellow sand	W/B	Workability (cm)
L-blond	1	3	—	—	0.758	15.0
L-black	1	—	3	—	0.800	14.5
L-yellow	1	—	—	3	0.750	14.8

Table 3.
Composition of trial mortar mixtures.

Composition	Compressive strength (MPa)	Porosity % (RILEM CPC11.3)
L-blond	1.14	26.42
L-black	1.03	27.11
L-yellow	0.98	30.79

Table 4.
Properties of the produced mortars at 28 days.

water absorption capacity of the sand grains, influence both the fresh (workability) and the hardened properties (porosity, strength) of the produced mortars.

The natural sands can be of similar origin with the crushed but weathering not only rounded the particles but also changed the proportions and removed most of the light minerals, such as the flaky micas. Due to these differences, mixtures with crushed sand often display higher water demand and lower workability than the corresponding composite with glaciofluvial sand. Additionally, crushed sand has a positive impact on long-term strength. It seems that, when rough-grained sand is used, strong cohesion with the binder can be achieved, as shown in **Figure 4**, where mortars with rounded and crushed sand were examined under scanning electron microscopy (SEM) [17].

The mechanical features, particle shape, grading, and physical properties, such as moisture absorption, sand equivalent value, are what can be labeled as properties of interest in the aggregates when used in mortars. Some of these most important properties are shown in **Table 5**.

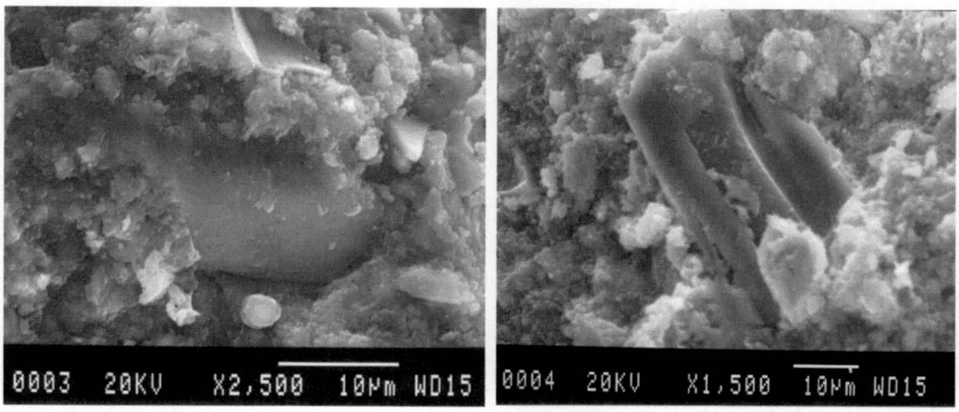

Figure 4.
SEM examination of rounded sand grain (left) where there is a gap in the contact zone and angular grain with strong cohesion (right).

Property	Regulation
Determination of rock compressive strength	ASTM C170
Determination of disintegration resistance (health) of aggregates (sodium sulfate method).	ASTM C88 AASHTO T104)
Determination of mineral hardness	the MOHS scale
Determination of specific gravity of aggregates	BS 812/ AASHTO T 19.
Determination of moisture absorption of aggregates.	AASHTO T85
Determination of granulometric analysis	AASHTO T27/ AASHTO T11/ EN933-1
Determination of ultra-fine crushed material by rinsing	A8TM 0117 (AASHTO T37)
Determination of equivalent sand	AASHTO T176
Determination of abrasion resistance of aggregates	BS 812/75
Determination of wear in aggregate crushing	BS 812/75
Determination of wear on the impact of aggregates	BS 812/75
Determination of plaque index	88,812/75 Section 105.1

Table 5.
Important properties of aggregates to be used in mortar production.

The bond behavior in the interface between the binder and the aggregates has a strong effect on the mortar properties since the effectiveness of the reinforcement provided by the addition of particles depends on the interfacial bond (**Figure 5**). This is since the size, shape, and content of the particles predominantly control the morphological features of the internal structure of the composite.

The test results showed that with increasing volume fraction of aggregate, the compressive strength of the composite decreases, which is different from the prediction of conventional composite theories. The possible explanation of this result is

Figure 5.
Macroscopic examination of contact zones of natural aggregates and binders in old mortars. Despite the presence of cracks in the binder in the left image, the cohesion is strong. On the right, there are pores on the interface probably due to the high content in aggregates in relation to the binder.

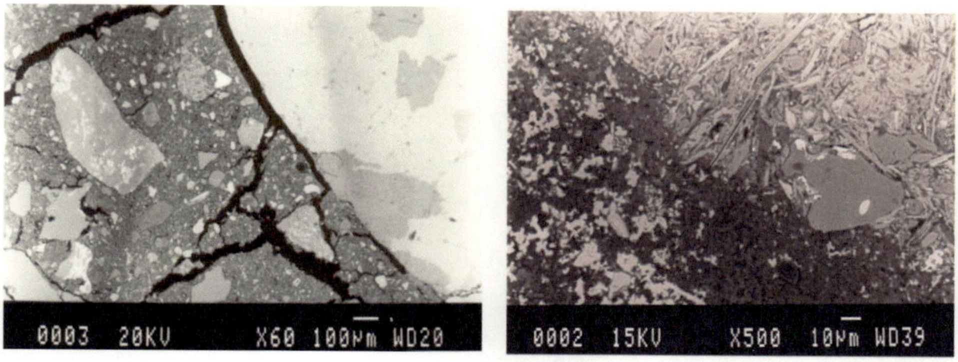

Figure 6.
Examination under SEM of natural aggregate and lime binder with weak ITZ (left) and crushed brick as aggregate and lime-pozzolan binder with strong ITZ (right).

based on the interface transition zone (ITZ) around the aggregate, which is the weak zone in composites (**Figure 6**) [15]. With more aggregate added into the mixtures, more interfaces are formed in the hardened material. The compatibility between the aggregate of the paste affects the development of strong cohesion at the aggregate-matrix interface in many cases and that usually indicates the good performance of the mortar. As aggregates are, by weight or by volume, the major component of mortars, they can be a source of silica, which can react in certain conditions with the binder, leading to the formation of reaction rims at the edge of the grains and recrystallization along with the pre-existing cracks (**Figure 7**).

Apart from the different types of aggregates as their mineralogy is concerned, the volume content in the mixture, the maximum size, and their gradation influences the structure of a binder—aggregate mixture [3, 5]. The analysis of mortars reveals that higher strength values are attained for lime mortars of low binder/aggregate (B/A) ratio (1:1.5, 1:2.5, and 1:3) which contained sand (0–4 mm). Coarse aggregates have

Figure 7.
Old mortars under the polarized microscope (x10). Reaction rim in the interfacial zone of the binders and the aggregates used.

Figure 8.
Pores and cracks in the structure of lime mortar with coarse aggregates (polarized microscope, x15).

Figure 9.
Cracks inside the binder where they meet the aggregate volume as the obstacle.

contributed positively to the volume stability of lime mortars. The microstructure has recorded the restriction of volume changes in cases where coarse aggregates have been used in the structure of lime mortars (**Figure 8**).

However, it is well recognized that coarse aggregate particles can act as crack arresters, as they restrict the shrinkage of the binder so that under an increasing load, extra energy is absorbed for the formation of a new crack (**Figure 9**) [36].

3. Designing repair mortars

Usually, a detailed analysis of the authentic building materials is performed to establish an opinion about the materials and techniques used during the construction phase [2]. Based on the results of this analysis, the design and laboratory production of some materials follows [14, 37]. The destructive consequences from the use of incompatible repair materials are related to different physical, chemical, and elastic characteristics that many new materials possess in relation to the old lime-based ones. For this reason, the quality of the materials used in intervention works is of primary importance for the longevity and economy of interventions. However, standard test methods and recommendations have not yet been developed despite the effort at the European level.

As river sand remains as one of the most widely used fine aggregates due to its desirable properties an increased tendency to use and it is observed. With an increase in construction activities, the demand for river sand has also been increasing. As a result, it has been mined at a high rate, depleting its natural resources and causing serious environmental issues. Also, owing to the excess cost of transportation, the natural river sand has become expensive. Hence, industries are shifting to other materials, such as crushed sand. But as the demand for building materials will continue to increase, their sources for crushed sand might also get exhausted. Therefore, there is a need to replace the fine aggregate either completely or partially with an alternative material that can satisfy the properties required for concrete, which is cost-effective and at the same time sustainable. Finding an alternative material to river sand has now become imperative.

The incentive to use sand from building demolition in repairing mortars derives from different needs. Natural sand originating from rivers is becoming rare, while the extraction of aggregates from quarries carries an increased administrative cost due to new strict legislations.

Both practices are not considered environmental friendly and, thus, the criteria and legislation for sand extraction are becoming stricter and demanding, while in some places, good quality natural sands are not available. On the other hand, the increased waste production offers the availability of large volumes of recycled materials and public concern about the environment pushes toward their utilization. The possibility of incorporating fine recycled sand originating from construction and demolition waste in lime-based traditional mortars. The study showed that the recycled sand had an even grain distribution, without any hazardous material and low content of soluble salts [38]. The mortars mixtures with recycled sand showed increased water demand and reduced workability compared to mortars with natural sands, even when superplasticizer was used [39]. The mechanical strength measured at 28 and 90 days showed good results as the mortars with lime and recycled sand had higher compressive strength compared to mortars with natural sands [40].

Additionally, several industrial wastes, (fly ash, demolition waste, slag, glass, brick waste, and plastic), have been shown to be suitable as construction materials and readily follow the design requirements. The substitution of the siliceous aggregate with plastic sand leads to a decrease in mechanical properties, opportunities in the use of these materials are not affected, especially for applications that do not require a structural function [41].

4. Conclusions

The mechanical and physical properties of a mortar both at fresh state, but also long-term, depend on multiple factors, including binder type, curing time, binder—aggregate and binder—water ratios, nature, shape, and grading of aggregates, the compaction degree, and also the environment in which they function. As mortars are composite materials, each component has a special role in the ultimate quality of the material. Aggregates, being of great volume in the mortar mass, significantly influence the structure and the properties achieved in all states of mortar production. The analyses of old mortars revealed the continued presence of sand in the mortars from pre-history up to the cement era. Coarser grains were also used in the technology of mortars. Generally, it is accepted that the strongest mortar mixes are produced from well-graded, clean, and angular aggregates. Usually, they were of local origin following principles of ecology and economy.

The same principles should be applied today having the technological evolution as an alley to protect the environment and work on the benefit of the constructions. Understanding the mechanisms of action and the parameters affecting significant properties in mortars, a well-engineered mixture can be achieved utilizing alternative solutions to protect natural resources and at the same time bring to the market high-quality innovative mortars. Recycled sands are promising materials in construction as after specific tests, they can be utilized either in repairing old structures or even in preparing new cement-based mortars.

Author details

Maria Stefanidou* and Parthena Koltsou
Laboratory of Building Materials, School of Civil Engineering, Aristotle University of Thessaloniki, Thessaloniki, Greece

*Address all correspondence to: stefan@civil.auth.gr

References

[1] Pachta V, Stefanidou M, Konopisi S, Papayianni I. Technological evolution of historic structural mortars. Journal of Civil Engineering and Architecture. 2014;**8**(7):846-854

[2] Papayianni I. Durability lessons from the study of the old mortars and concretes, P.K. Mehta Symposium on Durability of Concrete, Nice, May 23, 1994, Organized by CANMET/ACI, pp. 1-30

[3] Martin R. Palladius Traite' D'Agriculture. Livre I. Societe' D'Edition 'Les Belles Letres'; 1976

[4] Eichholz, Pliny the Elder. Natural History. London: William Heinemann Ltd., Harvard University Press; 1962. 2nd edition 1971

[5] Hicky Morgan M. Vitruvius: The Ten Books on Architecture. Cambridge: Harvard University Press; 1914

[6] Palivou K. Akrotiri of Thera: The Constructional Technology. Athens: Library of the Archaeological Society at Athens; 1999. (in Greek)

[7] Orlandos AK. The Building Materials of Ancient Greeks, According to Writers, Inscriptions and Monuments: Part A. Vol. 2. Athens: Ancient Greek Architecture; 1958. (in Greek)

[8] Bugini R, Salvatori A. Investigation of the characteristics and properties of "Cocciopesto" from the ancient Roman period. Conservation of Stone and Other Materials. 1991;**1**:386-393

[9] Papayianni I, Stefanidou M. Repair mortars for monuments in Byzantine architecture. In: Proceeding of 5th International Congress on Restoration of Architectural Heritage, CICOP, Firenze. 2000. pp. 1671-1683

[10] Baronio G, Binda L, Tedeschi C. Thick mortar joints in Byzantine buildings: Study of their composition and mechanical behaviour. In: Proceeding of International Conference on Studies in Ancient Structures, Istanbul. 1997. pp. 235-244

[11] Papayianni I, Stefanidou M. Repair mortars suitable for interventions of Ottoman monuments. In: Proceeding of International Conference on Studies in Ancient Structures, Istanbul. 1997. pp. 255-264

[12] Papayianni I. Technology of mortars and bricks used in Ottoman monuments in Thessaloniki. In: Arun G, Seçkin N, editors. Proceeding of International Conference on Studies in Ancient Structures, Istanbul. 1997. pp. 245-253

[13] Maravelaki-Kalaitzakia P, Bakolas A, Moropoulou A. Physico-chemical study of cretan ancient mortars. Cement and Concrete Research. 2003;**33**:651-661

[14] Papayianni I. Design of compatible repair materials for the restoration of monuments. International Journal for Restoration. 2004;**10**(6):623-636

[15] Lanas J, Alvarez-Galindo JI. Masonry repair lime-based mortars: factors affecting the mechanical behaviour. Cement and Concrete Research. 2003;**33**(11):1867-1876

[16] Stefanidou M. Crushed and river-origin sands used as aggregates in repair mortars. Geosciences. 2016;**6**(2):23. DOI: 10.3390/geosciences6020023

[17] Stefanidou M, Papayianni I. The role of aggregates on the structure

and properties of lime mortars cement and concrete composites. 2006;**27**(9-10):914-919

[18] Arulmoly B. Chaminda Konthesingha, Anura Nanayakkara Performance evaluation of cement mortar produced with manufactured sand and offshore sand as alternatives for river sand. Construction and Building Materials. 2021;**297**:123784

[19] Handbook of Materials Selection. Myer Kutz Associates editors. In: Chapter 12 Composite Materials, Carl Zweben. New York: John Wiley & Sons; 2002. pp. 357-386

[20] EN 13139:2002 Aggregates for mortar

[21] Shin K-J, Lee S-C, Kim YY. Role of fine aggregates on mechanical properties of mortar. Materials Research Innovations. 2015;**19**(8):S8-690-S8-692. DOI: 10.1179/1432891715Z.0000000001778

[22] Santos AR, do Rosário Veiga M, Silva AS, de Brito J, Álvarez JI. Evolution of the microstructure of lime-based mortars and influence on the mechanical behaviour: The role of the aggregates. Construction and Building Materials. 2018;**187**:907-922

[23] Papayianni I, Stefanidou M, Christodoulou S. Influence of the sand fines on the mechanical and physical properties of lime-based renders and plasters. In: Válek J, Groot C, Hughes JJ, editors. Proceedings of the 2nd Conference and of the Final Workshop of RILEM TC 203-RHM, Prague. 2010. pp. 1135-1144

[24] Westerholm M, Lagerblad B, Silfwerbrand J, Forssberg Influence E. of fine aggregate characteristics on the rheological properties of mortars. Cement and Concrete Composites. 2008;**30**(4):274-282

[25] EN1015-1:1999 Methods of test for mortar for masonry Part 1: Determination of particle size distribution (by sieve analysis)

[26] Amparano FE, Xi Y, Roh Y-S. Experimental study on the effect of aggregate content on fracture behavior of concrete. Engineering Fracture Mechanics. 2000;**67**:65-84

[27] Asamoto S, Ishida T, Maekawa K. Volumetric stability of aggregates and shrinkage of concrete as composites. Journal of Advanced Concrete Technology. 2008;**6**(1):77-90

[28] Chang TP, Taso KL, Lin BR. Effect of aggregate on fracture properties of high-performance concrete. In: Mihashi H, Rokugo K, editors. Fracture mechanics of concrete structures, Proc. of FRAMCOS-3. D-79104 Freiburg: Aedificatio Publishers, 1998. pp. 151-160

[29] Neville AM. Properties of Concrete. Fourth ed. New York: John Wiley & Sons Inc.; 1996. pp. 244-248

[30] Zhang W, Zakaria M, Hama Y. Influence of aggregate materials characteristics on the drying shrinkage properties of mortar and concrete. Construction and Building Materials. 2013;**49**:500-510

[31] Stefanidou M, Anastasiou E, Filikas KG. Recycled sand in lime-based mortars. Waste Management. 2014;**34**:2595-2602

[32] Zhang J, An X, Nie D. Effect of fine aggregate characteristics on the thresholds of self-compacting

paste rheological properties. Construction and Building Materials. 2016;**116**(30):355-365

[33] Pavıa S, Toomey B. Influence of the aggregate quality on the physical properties of natural feebly-hydraulic lime mortars. Materials and Structures. 2008;**41**:559-569. DOI: 10.1617/s11527-007-9267-4

[34] EN 459-2:2010 Building lime – Part 2: Test methods

[35] EN1014-3 Part 3:1999 Determination of consistence of fresh mortar (by flow table)

[36] Lea FM. The Chemistry of Cement and Concrete. London: Edward Arnold Ltd; 1970. pp. 406-559

[37] Moropoulou A, Bakolas A, Moundoulas P, Aggelakopoulou E, Anagnostopoulou S. Strength development and lime reaction in mortars for repairing historic masonries. Cement and Concrete Composites. 2005;**27**(2):289-294

[38] Martínez I, Etxeberria M, Pavón E, Díaz N. A comparative analysis of the properties of recycled and natural aggregate in masonry mortars. Construction and Building Materials. 2013;**49**:384-392

[39] Anastasiou E, Georgiadis-Filikas K, Stefanidou M. Utilization of fine recycled aggregates in concrete with fly ash and steel slag. Construction and Building Materials. 2014;**50**:154-161

[40] Jiménez JR, Ayuso J, López M, Fernández JM, de Brito J. Use of fine recycled aggregates from ceramic waste in masonry mortar manufacturing. Construction and Building Materials. 2013;**40**:679-690

[41] Iucolano F, Liguori B, Caputo D, Colangelo F, Cioffi R. Recycled plastic aggregate in mortars composition: Effect on physical and mechanical properties. Materials & Design. 2013;**52**:916-922

Chapter 2

Bentonite Clay Modified Concrete

Metta Achyutha Kumar Reddy and Veerendrakumar C. Khed

Abstract

Replacing cement with pozzolanic materials to some extent in construction is found to be one of the sustainable approaches in the construction industry. Pozzolanic materials of industrial origin like fly ash and Ground Granulated Blast furnace Slag will have to be replaced with natural pozzolanic materials once the world moves towards renewable energy sources. Bentonite is one such pozzolanic clay material that is rich in SiO_2 content. A little research was made to assess the performance of bentonite modified concrete. Based on those, an improvement in the fresh, hardened, durability properties was reported. This chapter presents the current scenario on the development of bentonite modified concrete. It also reviews the literature about the physical & chemical properties of bentonite, bentonite blended cement mortar, bentonite modified cement concrete, and reinforced concrete. The history and development of Bentonite modified concrete were also briefly presented in this chapter.

Keywords: bentonite clay, bentocrete, workability, strength, durability

1. Introduction

A developing country like India must cater to the construction activity of buildings, bridges, transport facilities, industrial units, and many more on a vast scale. The spiraling costs of such building materials as cement and reinforcing steel hurt the development activity [1].

Cement and construction industries were generating 7–8% of CO_2 emissions globally by manufacturing cement and production of concrete [2]. The utilization of industrial wastes as pozzolanic materials in concrete to enhance workability, strength, and durability emissions became a trend since the early 1980s [3]. The generation of industrial wastes would not have happened if the industries had been shut down [4]. The Discovery of alternatives to the pozzolanic materials generated from industrial wastes was needed. Bentonite is clay, contains a rich amount of SiO_2 [5]. It could be used natural pozzolanic material instead of industrial wastes if the generation was shut down permanently [6]. There have been a few experiments in the past to see if bentonite could be used in concrete. The investigations began in Pakistan and were later expanded by several countries [7, 8]. The use of bentonite in concrete was shown to have favourable benefits in terms of strength and durability [9, 10].

This chapter presents a review of available and related published literature on the properties of bentonite. It also reviews the literature about the physical & chemical properties of bentonite, bentonite blended cement mortar, bentonite modified

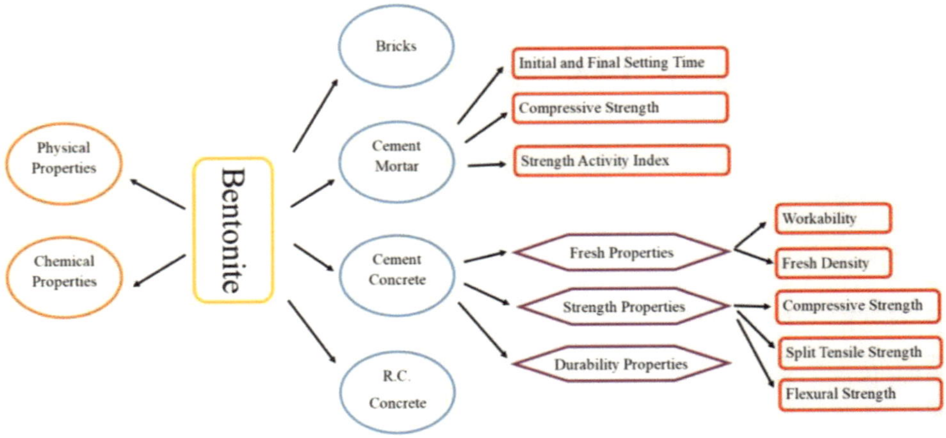

Figure 1.
Overview of the chapter.

cement concrete, and reinforced concrete. The history and development of Bentonite modified concrete were also briefly reviewed in this chapter, **Figure 1** shows the overview.

2. Bentonite

Bentonite is an earth mineral that shows expanding conduct and follows Pozzolanic properties [11, 12]. Huge amounts of bentonite assets are accessible everywhere. It has huge applications in penetrating liquids, foundries, glues, fading earth, earthenware production, etc. [13]. Mirza Initiated the exploration work by involving bentonite as a substitute material to solidify in concrete in 2009 [14]. A lot of examinations were done by scientists from various areas all through the globe. They announced a few realities connected with bentonite use because of the test results [15]. Most of the authors utilized Pakistani bentonite accessible from various areas of Pakistan. Mirza et al. [14] incorporated Karak bentonite in concrete, investigated the properties of concrete. Ahmad et al. [16], employed Jahangira bentonite in concrete and evaluated the properties of cement mortar. Afzal et al. [17] utilized Khyber bentonite in concrete assayed the autogenous shrinkage strain. Lima-Guerra et al. [18] developed concrete using bentonite from the Amazon region of Brazil. Production of artificially expanded clay aggregates is also possible by making use of bentonite [19].

2.1 Physical properties

Bentonite is for the most part accessible in different shadings and structures. A Few creators announced with regards to the shading. In their exploratory work, Mirza et al. [14] used greenish dim and searing green bentonite while light yellow-shaded bentonite was of bentonite used Ahmad et al. [16]. **Table 1** shows the outline of the actual properties of bentonite detailed by certain authors. Practically all creators revealed various qualities. The explanation for this trait is to change in the area of the source.

S. No.	Property	[10]	[14]	[16]	[20]
1	Specific gravity	2.82	2.63	2.44	2.79
2	Average particle size	4.75 µm	17% retained on # 325 mesh	13.5% retained on # 325 mesh	4.32 µm
3	Blains fineness (cm2/gm)	4800	—	2689	4800

Table 1.
Physical properties of bentonite.

	Property	[10]	[9]	[11]	[12]	[14]	[15]
1	Silicon dioxide	54.55	49.44	65	49.63	52.1	51.11
2	Aluminum Oxide	20.19	19.7	15	21.11	13.4	16.83
3	Ferric Oxide	8.60	6.2	3	3.23	7.5	7.65
4	Magnesium Oxide	4.20	1.61	6.5	12.56	2.64	7.57
5	Calcium Oxide	7.28	7.45	2.66	3.59	12.0	6.60
6	Sodium Oxide	1.27	0.87	0.12	0.44	-	0.29
7	Potassium Oxide	3.92	0.63	0.27	2.09	2.64	1.34
8	Phosphorus pentoxide	1.107	-	-	0.11	-	0.29
9	Titanium Oxide	0.91	-	-	0.49	-	1.29
10	Loss on Ignition	5.42	13.74	6	-	8.61	6.75

Table 2.
Chemical properties of bentonite.

2.2 Chemical properties

The chemical composition shows a huge effect on the mechanical properties of concrete. **Table 2**. displays the chemical properties of bentonite. The studies reveal a larger amount of SiO_2 presence, $Al2O_3$ as a second focal component in bentonite. It prompts the occurrence of a pozzolanic reaction during the hydration process. Most of the researchers revealed that those are inside limits. 13.74 LoI esteem was accounted for by Mirza et al. [14], which is the only value not within the allowable limits.

3. Bentonite modified cement mortar

The bentonite's behavior in concrete mortars like consistency, introductory and last setting times, strength activity, and compressive strength was accounted for in a few investigations.

3.1 Normal consistency

Bentonite shows 75% as consistency value while Ordinary Portland Cement (OPC) accomplishes at 30–35%, 21 percentage bentonite-OPC combination achieves 35% consistency was accounted for by Memon et al. [10]. Nonetheless, it has been seen that the consistency is straightforwardly relative to how much bentonite is added [21].

3.2 Initial and final setting time

Generally, Initial & final setting time tests will be performed as per standard procedure IS 4031. Upon testing, bentonite displays 68 and 190 minutes where shown OPC 43 and 125 minutes after expansion with water [15]. Beginning and last setting times expanding by expansion bentonite to solidify. The expansion of bentonite to the concrete mortar upgrades the yield pressure of new concrete mortar [22].

3.3 Strength activity index

Mirza et al. [14] directed a few tests on strength activity index according to standard method ASTM C618 for bentonite blends at various temperatures, shown in **Figure 2**. A more prominent strength activity index was seen in 150°C warmed bentonite contrasted and various temperatures. Ahmad et al. [16] revealed 87.3 and 85.23% SI for bentonite at 7 years old and 28 days. 3% bentonite-concrete blends show better strength activity index at 7days, remaining stirs up to 21% displays best similarly among all mixes. A slight decrement in the strength action was seen by bentonite joining in concrete mortars [23].

3.4 Compressive strength

Mirza et al. [14] performed tests as per ASTM C109, which announced that 150°C warmed 20% bentonite-concrete blends accomplish great compressive strength at 7 years old and 28-days remaining blends shown lower than control concrete mortar, displayed in **Figure 3** [9]. Ahmad et al. [16] led probes bentonite blends at room temperature and 500°C; all blends have shown lower compressive strength than the control blend. The concrete mortar containing 20% bentonite shows more compressive strength than control concrete mortar [24]. 30% bentonite-concrete blends show

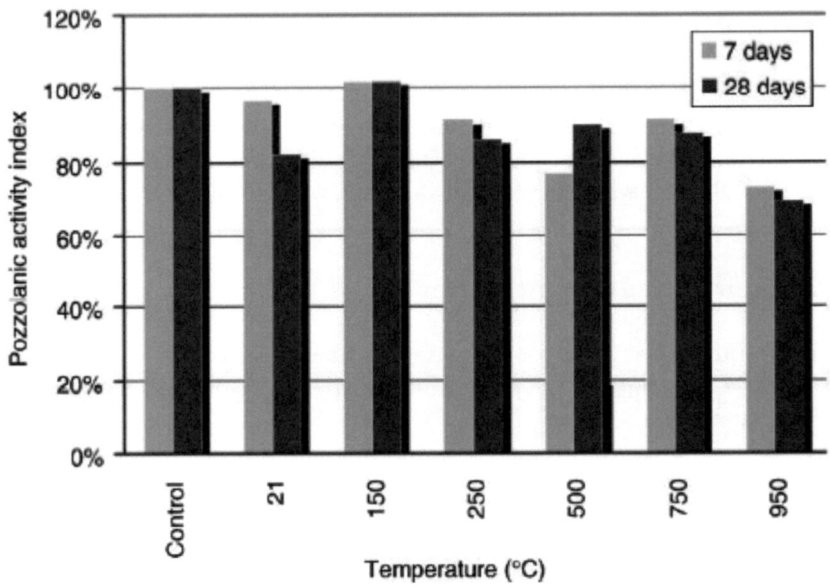

Figure 2.
Strength activity Indices of different bentonite mixes [14].

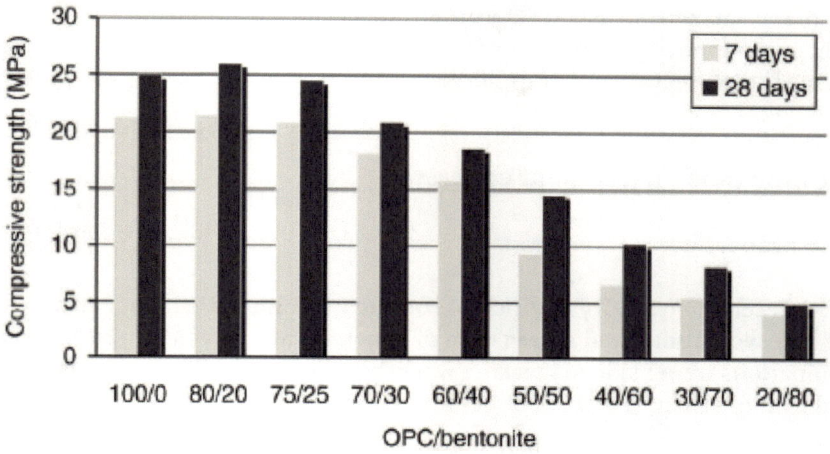

Figure 3.
Compressive strength of different bentonite mixes [14].

higher protection from $MgSO_4$ and Na_2SO_4 [11]. Reddy et al. [25] revealed that 20% bentonite-concrete blend showed ideal compressive strength following 3, 7, and 28 days relieving among all mixes [25]. Bentonite adjusted concrete mortar shows lower compressive strength than control concrete mortar in adjustment sewage slime [26]. Better compressive strength was shown by bentonite adjusted concrete mortar in the redesign of adobe structures over metakaolin [27].

4. Bentonite clay bricks

An attempt was made to manufacture eco-friendly bricks using a combination of bentonite, clay, and lime. Bentonite content was considered as a variable, the properties of the brick were essayed. 31.91% of the carbon footprint was reduced by 20% of

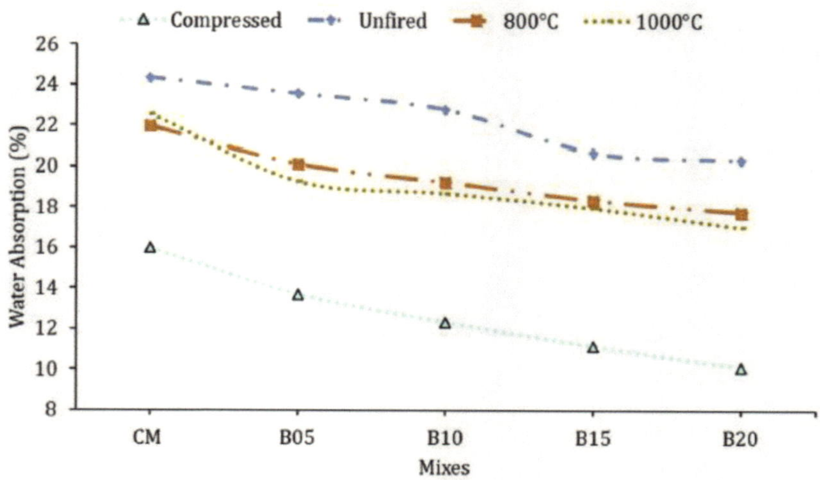

Figure 4.
Water absorption of clay, unfired and fired bricks [28].

bentonite replacement with clay. The durability of the bricks enhances by the incorporation of bentonite in bricks manufacturing. Water absorption was also reduced by improving the percentage of bentonite substitution, displayed in **Figure 4** [28].

5. Bentonite modified cement concrete

5.1 Fresh properties

1–2% on cement quantity addition of sodium bentonite to cement concrete will improve the workability and segregation of concrete can be avoided [7]. Later, plenty of examinations were made to measure workability precisely. The elastic modulus was reduced by more than 15 percent of bentonite for the plastic concrete [29].

5.1.1 Workability

A radical lessening in workability was seen at higher rates (least of 20%) of bentonite replacement in concrete. Bentonite usage was done in bring down rates (0–21 at 3% stretch) by Memon et al. [10]; he prescribed utilizing superplasticizer expected to upgrade workability, slump values were displayed in **Figure 5**. The addition of a superplasticizer can attain the desired workability. Production of bentonite modified concrete is also possible by using recycled crushed aggregates [30]. The concrete made with crumb rubber and bentonite also exhibits decrement in workability [31].

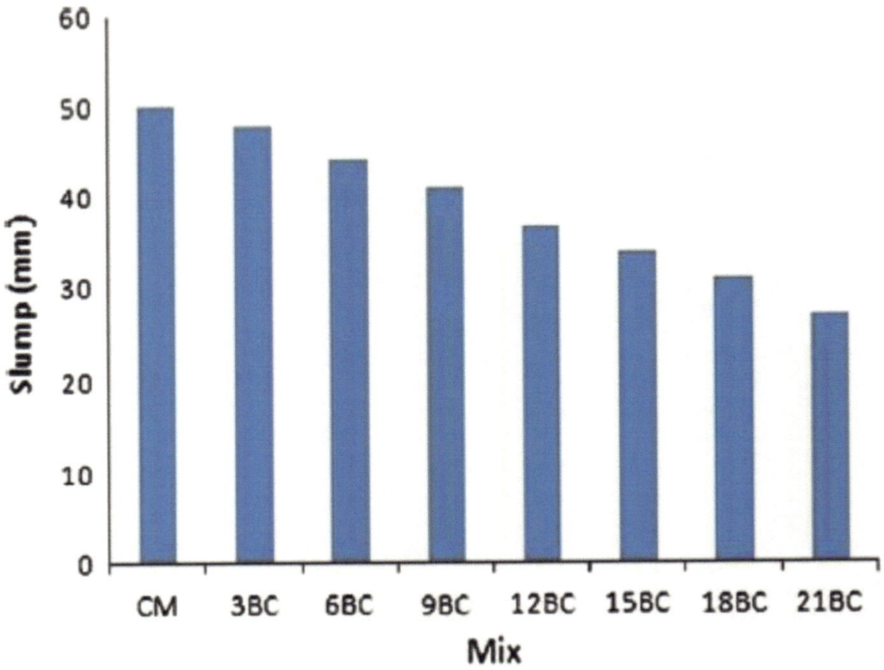

Figure 5.
Slump values of all mixes [10].

5.1.2 Fresh concrete density

The fresh concrete density was decreased upon increasing the bentonite content in concrete. The density values are displayed in **Figure 6**. It is attributed to the lower specific gravity of bentonite, examined by standard procedure ASTM C642 [10].

5.2 Hardened properties

A lot of studies detailed the impact of bentonite on the mechanical properties of hardened concrete. Compressive strength was measured at various periods of concrete. Resistance of cement was tried and detailed under different climatic conditions. Most studies revealed that lower qualities were seen by the addition of bentonite at early ages (3,7 and 28 days), the better exhibition displayed at later ages (56 and 90 days) of restoring similarly with control concrete.

5.2.1 Compressive strength

Mirza et al. [14] conducted investigations by altering bentonite (0–50%) at room temperature and warmed at 150°C for 3 hours. Ahmad et al. [16] explore bentonite warmed at 500°C. They announced that 20% of bentonite (warmed at 150°C for 3 hours) subbed substantial blends display better compressive strength at 28 days restoring among all. The compressive strength of cement is straightforwardly corresponding to the temperature at which bentonite was warmed for 3 hours. Khushnood et al. [20] announced warmed bentonite mixed substantial blends showed higher compressive strength than crude bentonite. Memon et al. [10] analyzed up to 21% bentonite replacement, revealed that lower compressive strength shows at 3 years old, 7day, better compressive strength displayed at 28, 56 days in the wake

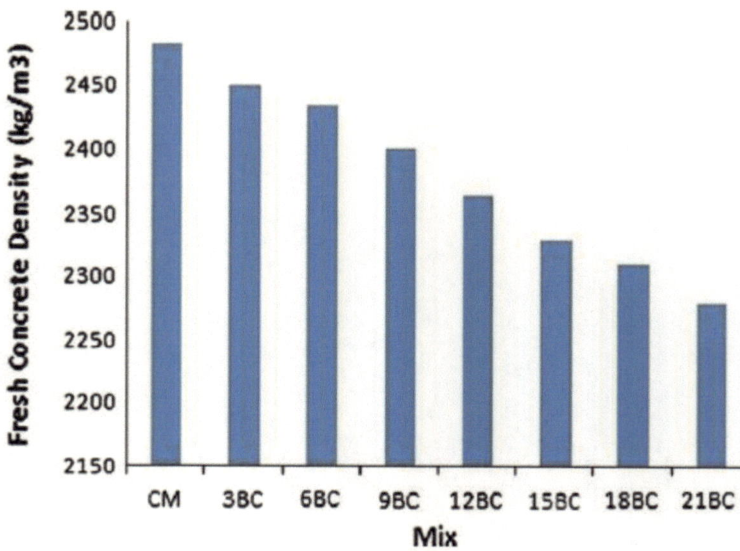

Figure 6.
The fresh concrete density of all mixes [10].

of relieving, shown in **Figure 7**. Akbar et al. [32] researched at 20% bentonite replacement, poor compressive strength results were accounted for while contrasting and control blend.

Reddy et al. [25] utilized 10–30% bentonite replacement at 5% stretches. Tests were led, detailed that lower compressive strength was noticed for mixed blends among all. It was seen that 30% of bentonite substitution brings about higher compressive strength than control mix [3, 33]. Divyana [34] and Kadar and Dhanalakshmi [35] detailed that 20% bentonite replacement showed more noteworthy compressive strength than control blend [34, 36]. Chamundeeswari [37] and Selvaraj and Priyanka [38] announced that bentonite expansion diminishes in compressive strength of cement. Chandrakanth et al. proposed that 5% bentonite expansion accomplishes the greatest compressive solidarity to the substantial [39]. Aravindhraj and Sapna proposed that a 15-half bentonite-quarry dust combination shows the most extreme compressive strength at 28 days [40].

Shabab et al. [5] examined bentonite-fly ash mixes, stated that the concrete contains an equivalent combination (50–50%) of bentonite and fly ash displays improved outcomes at the age of 90 days, displayed in **Figure 8** [5, 41]. Bentonite modified concrete has shown better compressive strength in lateral days even though the performance was poor in the early days [32, 42]. Bentonite replacement can be done up to 10% in the making of concrete combined with waste rubber tires [43]. The addition of bentonite in the concrete showed a better result than wheat straw ash (WSA) as a supplementary cementitious material, shown in **Figure 9** [44]. An optimal bentonite substitution percentage was found as 2.7 while using the combination of kaolin and slag [45]. In foamed concrete, it was observed that more than 10% bentonite replacement would result from the decrement in compressive strength, and fluidity can be improved by more than 20% bentonite substitution [46].

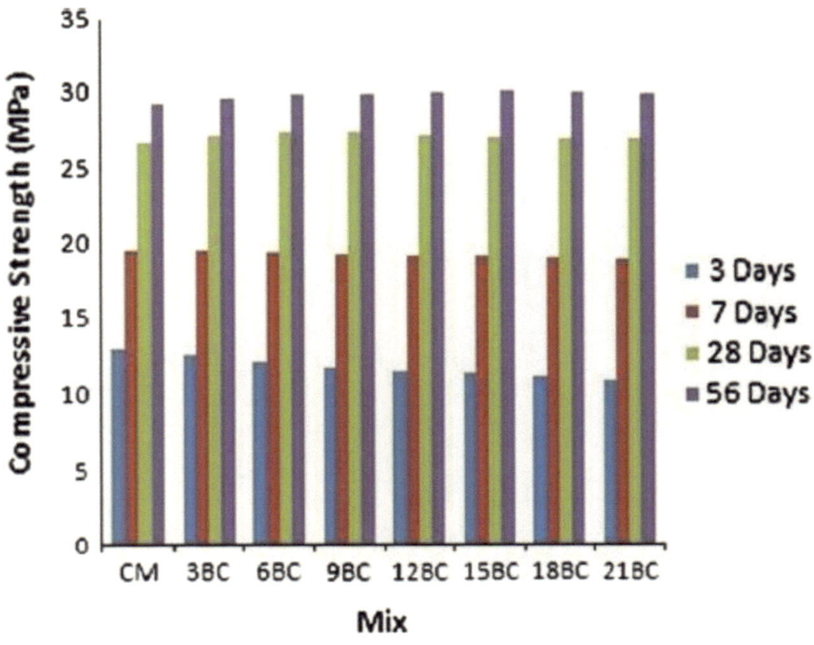

Figure 7.
Compressive strength of bentonite mixes [10].

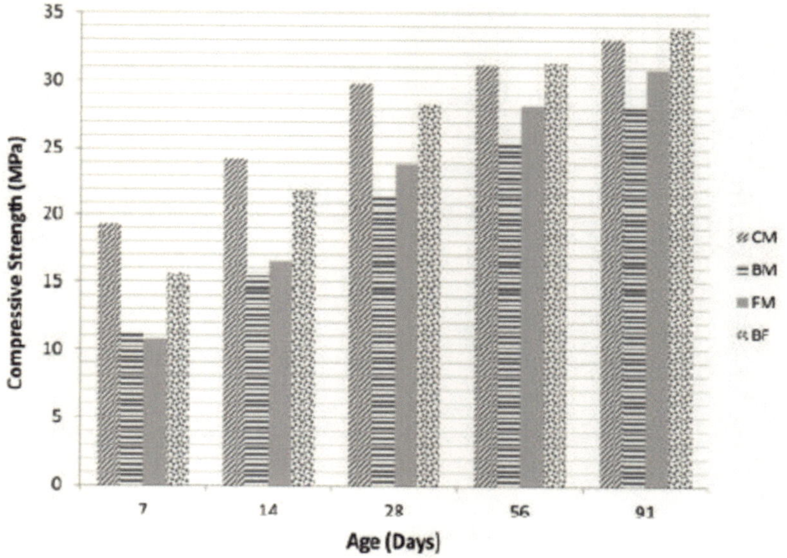

Figure 8.
Compressive strength of bentonite-fly ash mixes [5].

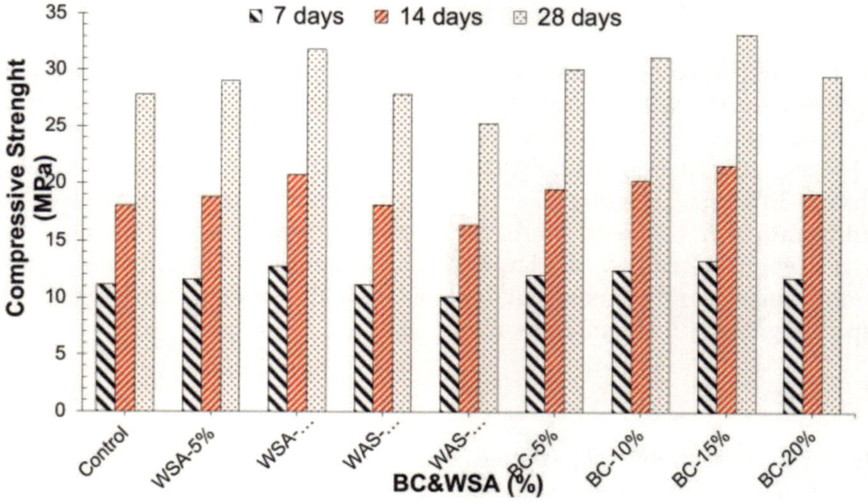

Figure 9.
Compressive strength results of WSA & bentonite clay concrete [44].

5.2.2 Split tensile strength

Karthikeyan et al. [47] detailed that 30% replacement of bentonite brings about the most extreme split tensile strength to concrete, showed in **Figure 10**. Reddy et al. [25] led tests according to standard technique IS5816, revealed that lower split tensile strength was seen in bentonite mixed blends. Divyana [34] revealed that better-parted rigidity was accomplished by 20% bentonite substitution. Mohammed et al. [36] tried on bentonite-steel slag blend with different extents, detailed that 20–60% bentonite-steel slag mixtures [35]. Chandrakanth et al. [39] revealed that 5% bentonite expansion further develops results split tensile strength of cement.

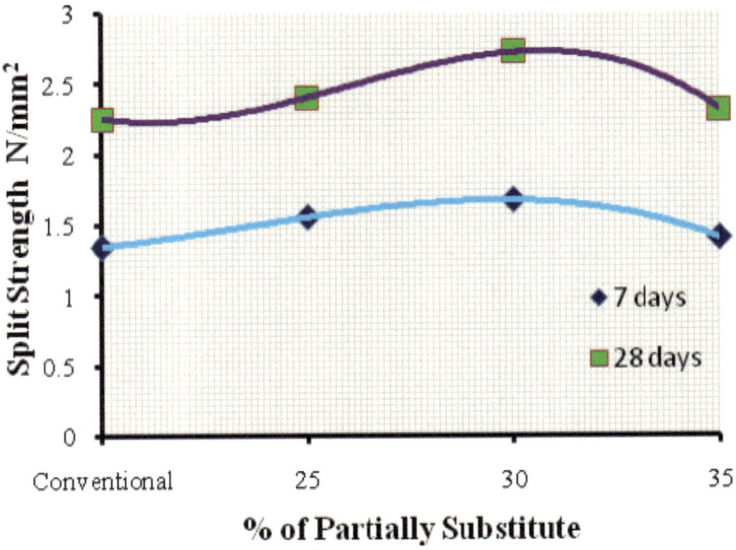

Figure 10.
Split tensile strength of bentonite mixes [47].

5.2.3 Flexural strength

Mirza et al. [14] and Khushnood et al. [20] detailed that no change in flexural strength was seen by warmed bentonite over crude bentonite, a decline in flexural strength was accounted for by the expansion of bentonite to concrete. Karthikeyan et al. [47] announced a slight augmentation in flexural strength upon bentonite expansion. 30% replacement displays the greatest flexural strength among all blends. Chandrakanth et al. [39] revealed that 5% of bentonite expansion works on the flexural strength of cement. All a large portion of the creators detailed no impact of bentonite replacement. Memon et al. [10] detailed a diminishing in water assimilation by expanding the level of bentonite mixing.

5.3 Durability

Durability shows a pivotal job when cement is presented to outrageous conditions [48]. A couple of examinations were done on the toughness of bentonite mixed cement-like sulfate attack, corrosive attack, and salt attack. Memon et al. [10] tried bentonite mixed cement (0–21% @3 span) against hydrochloric corrosive and Sulfuric corrosive. They revealed that obstruction is straightforwardly corresponding to bentonite mixing (up to 21%). Mirza et al. [14] concentrated on the impact of bentonite (20–80% @20intervel). Ahmad et al. [16] inspected structure 20–50% @10intervel against 2% $MgSO_4$ and 5%Na_2SO_4. They announced that substantial obstruction is expanded against sulfate by adding bentonite around 20–30%. Sreeniva et al. detailed that 15% bentonite mixed cement showed superb opposition against hydrochloric acid [49]. They also recommend that bentonite use results decline in opposition against NaOH. Swarup et al. revealed that the bentonite-fly debris (50–50%) mixed cement showed helpless obstruction against hydrochloric corrosive and NaOH [50].

6. Bentonite modified reinforced cement concrete

Mirza et al. [14] played out a test on the R.C.C beam having a design load of around 40 kN with a three-point loading, load versus deflection plots shown in **Figure 11**. R.C.C beams were cast by using 20% bentonite, 25&40% (warmed at 150°C for 3 hours) bentonite with OPC. They detailed those lower ultimate strengths were seen in bentonite mixed cement beams.

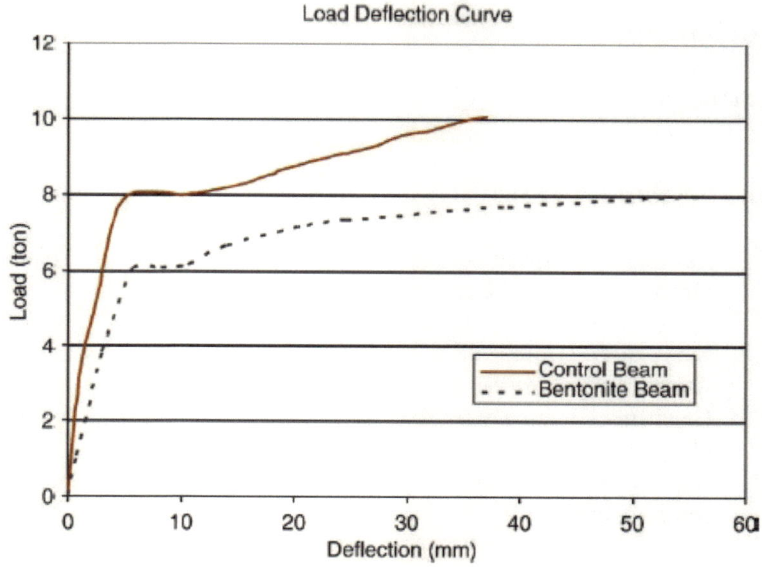

Figure 11.
Load vs deflection curve of concrete and bentonite modified concrete.

7. Conclusions

In this chapter, the current scenario on the development of bentonite modified concrete was presented. It also reviews the literature about the physical & chemical properties of bentonite, bentonite blended cement mortar, bentonite modified cement concrete, and reinforced concrete. The history and development of Bentonite modified concrete were also briefly presented.

8. Further research and recommendations

Based on the discussions, additional areas of research are identified to increase the understanding the' behavior. of bentonite modified concrete These areas of research include:

- Durability aspects of reinforced bentonite modified concrete columns.
- Fire resistance of bentonite modified concrete columns.

- Improvement of workability by adding suitable admixtures
- Development of bentonite-based geopolymer concrete.
- Effect of sustained loads.
- Effect of loading rate.

Conflict of interest

The authors declare no conflict of interest.

Acronyms and abbreviations

BC	Bentonite modified concrete
OPC	Ordinary Portland Cement
SEM	Scanning Electron Microscope
XRD	X-ray diffraction
ASTM	American Society Testing Method
C-S-H	Calcium-Silicate-Hydrate
MPa	Mega Pascal
kN	Kilo Newton
IS	Indian Standard

Author details

Metta Achyutha Kumar Reddy[1*] and Veerendrakumar C. Khed[2]

1 Koneru Lakshmaiah Education Foundation, Vaddesewaram, India

2 KLE Dr. M.S. Sheshgiri College of Engineering and Technology, Belgaum, India

*Address all correspondence to: achyuthakumarreddy@gmail.com

References

[1] Zeng Q, Li K, Fen-Chong T, Dangla P. Determination of cement hydration and pozzolanic reaction extents for fly-ash cement pastes. Construction and Building Materials. 2012;**27**(1):560-569. DOI: 10.1016/j.conbuildmat.2011.07.007

[2] Heikal M, Eldidamony H. Pozzolanic activity of fly ash. Silicates Industriels. 2003;**68**(9):111-117

[3] Siddique R. Utilization of industrial by-products in concrete. Procedia Engineering. 2014;**95**:335-347. DOI: 10.1016/j.proeng.2014.12.192

[4] Nizar K, Kamarudin H, Idris MS. Physical, chemical & mineralogical properties of fly-ash. Journal of Nuclear and Related Technology. Special Edition. 2007;**4**:47-51

[5] Shabab ME, Shahzada K, Gencturk B, Ashraf M, Fahad M. Synergistic effect of fly ash and bentonite as partial replacement of cement in mass concrete. KSCE Journal of Civil Engineering. 2016;**20**(5):1987-1995. DOI: 10.1007/s12205-015-0166-x

[6] Latawiec R, Woyciechowski P, Kowalski K. Sustainable Concrete Performance—CO2-Emission. Environments. 2018;**5**(2):27. DOI: 10.3390/environments5020027

[7] Murray HH. Chapter 6 bentonite applications. Developments in Clay Science. Applied Clay Mineralogy. 2006:111-130. DOI: 10.1016/S1572-4352(06)02006-X

[8] Indian Bureau of Mines, Indian Minerals Yearbook 2013 (Part-III: Mineral Reviews) 52 nd Edition Bentonite (Advance release) Government of India Ministry of mines Indian bureau of mines. 2015

[9] Reddy MAK, Rao VR. Utilization of bentonite in concrete: A review. International Journal of Recent Technology and Engineering. 2019;**7**(6C2):1-7

[10] Memon SA, Arsalan R, Khan S, Lo TY. Utilization of Pakistani bentonite as partial replacement of cement in concrete. Construction and Building Materials. 2012;**30**:237-242. DOI: 10.1016/j.conbuildmat.2011.11.021

[11] Elsayed AA, Amer N. Influence of bentonite content on the compressibility parameters of processed sand-bentonite mixtures. Concrete Research Letters. 2015;**6**(March):40-53

[12] Pomakhina E, Daniele D, Gaillot AC, Paris M, Ouvrard G. 29Si solid-state NMR investigation of pozzolanic reaction occurring in lime-treated Ca-bentonite. Cement and Concrete Research. 2012;**42**(4):626-632. DOI: 10.1016/j.cemconres.2012.01.008

[13] Saba S, Barnichon JD, Cui YJ, Tang AM, Delage P. Microstructure and anisotropic swelling behavior of compacted bentonite/sand mixture. Journal of Rock Mechanics and Geotechnical Engineering. 2014;**6**(2):126-132. DOI: 10.1016/j.jrmge.2014.01.006

[14] Mirza J, Riaz M, Naseer A, Rehman F, Khan AN, Ali Q. Pakistani bentonite in mortars and concrete as low-cost construction material. Applied Clay Science. 2009;**45**(4):220-226. DOI: 10.1016/j.clay.2009.06.011

[15] Nagy NM, Kónya J. Chloride ion migration in natural bentonite. Journal of Radioanalytical and Nuclear Chemistry.

2013;**298**(3):1519-1526. DOI: 10.1007/s10967-013-2682-9

[16] Ahmad S, Barbhuiya SA, Elahi A. Effect of Pakistani bentonite on properties of mortar and concrete. Clay Miner. 2011;**46**:85-92. DOI: 10.1180/claymin.2011.046.1.85

[17] Afzal S, Shahzada K, Fahad M, Saeed S, Ashraf M. Assessment of early-age autogenous shrinkage strains in concrete using bentonite clay as internal curing technique. Construction and Building Materials. 2014;**66**:403-409. DOI: 10.1016/j.conbuildmat.2014.05.051

[18] Lima-Guerra DJ, Mello I, Resende R, Silva R. Use of bentonite and organobentonite as alternatives of partial substitution of cement in concrete manufacturing. International Journal of Concrete Structures and Materials. 2014;**8**(1):15-26. DOI: 10.1007/s40069-013-0066-8

[19] Abdelfattah M, Kocserha I, Géber R, Tihtih M, Móricz F. Evaluating the properties and mineral phases of the expanded clay aggregates with the bentonite additive material. Journal of Physics: Conference Series. 2020;**1527**(1). DOI: 10.1088/1742-6596/1527/1/012030

[20] Khushnood RA, Rizwan SA, Memon SA, Tulliani JM, Ferro GA. Experimental investigation on use of wheat straw ash and bentonite in self-compacting cementitious system. Advances in Materials Science and Engineering. 2014;**2014**:1-12. DOI: 10.1155/2014/832508

[21] Al-Hammond AA, Frayyeh QJ, Abbas WA. Raw bentonite as supplementary cementitious material—A review. Journal of Physics: Conference Series. 1795;**1**:2021. DOI: 10.1088/1742-6596/1795/1/012018

[22] Kaci A, Chaouche M, Andréani PA. Influence of bentonite clay on the rheological behavior of fresh mortars. Cement and Concrete Research. 2011;**41**(4):373-379. DOI: 10.1016/j.cemconres.2011.01.002

[23] Mesboua N, Benyounes K, Benmounah A. Study of the impact of bentonite on the physico-mechanical and flow properties of cement grout. Cogent Engineering. 2018;**5**(1):1-12. DOI: 10.1080/23311916.2018.1446252

[24] Man X, Aminul Haque M, Chen B. Engineering properties and microstructure analysis of magnesium phosphate cement mortar containing bentonite clay. Construction and Building Materials. 2019;**227**:116656. DOI: 10.1016/j.conbuildmat.2019.08.037

[25] Reddy G, Rao VR, Reddy MAK. Experimental investigation of strength parameters of cement and concrete by partial replacement of cement with Indian calcium bentonite. International Journal of Civil Engineering and Technology. 2017;**8**(1):512-518

[26] Katsioti M, Katsiotis N, Rouni G, Bakirtzis D, Loizidou M. The effect of bentonite/cement mortar for the stabilization/solidification of sewage sludge containing heavy metals. Cement and Concrete Composites. 2008;**30**(10):1013-1019. DOI: 10.1016/j.cemconcomp.2008.03.001

[27] Andrejkovičová S, Alves C, Velosa A, Rocha F. Bentonite as a natural additive for lime and lime-metakaolin mortars used for restoration of adobe buildings. Cement and Concrete Composites. 2015;**60**:99-110. DOI: 10.1016/j.cemconcomp.2015.04.005

[28] Javed U, Khushnood RA, Memon SA, Jalal FE, Zafar MS. Sustainable incorporation of lime-bentonite clay composite for production of eco-friendly

bricks. Journal of Cleaner Production. 2020;**263**:121469. DOI: 10.1016/j.jclepro.2020.121469

[29] Kazemian S, Ghareh S, Torkanloo L. To investigation of plastic concrete bentonite changes on it's physical properties. Procedia Engineering. 2016;**145**:1080-1087. DOI: 10.1016/j.proeng.2016.04.140

[30] Masood B, Elahi A, Barbhuiya S, Ali B. Mechanical and durability performance of recycled aggregate concrete incorporating low calcium bentonite. Construction and Building Materials. 2020;**237**:117760. DOI: 10.1016/j.conbuildmat.2019.117760

[31] Adeboje AO, Kupolati WK, Sadiku ER, Ndambuki JM, Kambole C. Experimental investigation of modified bentonite clay-crumb rubber concrete. Construction and Building Materials. 2020;**233**:117187. DOI: 10.1016/j.conbuildmat.2019.117187

[32] Akbar J, Alam B, Ashraf M, Afzal S, Ahmad A, Shahzada K. Evaluating the effect of bentonite on strength and durability of high performance concrete. International Journal of Advanced Structures and Geotechnical Engineering. 2013;**02**(01):1-5

[33] Wei J, Gencturk B. Hydration of ternary Portland cement blends containing metakaolin and sodium bentonite. Cem. Concr. Res. 2019;**123**(May):105772. DOI: 10.1016/j.cemconres.2019.05.017

[34] Divyana R. An Experimental Study on Concrete Using Bentonite and Steel Slag. In: NCRACCESS-2015. 2015. pp. 14-18 Available from: www.internationaljournalssrg.org

[35] Kadar JMA, Dhanalakshmi G. Experimental investigation on concrete by partial replacement on cement by bentonite and coarse aggregate by steel slag. International Journal of Innovative Research in Science, Engineering and Technology. 2016;**5**(6):10302-10309. DOI: 10.15680/IJIRSET.2015.0506148

[36] Mohammed J, Kadar A, Dhanalakshmi G. Experimental investigation on concrete by partial replacement on cement by bentonite and coarse aggregate by steel slag. International Journal of Innovative Research in Science, Engineering and Technology. 2007;**3297**(6):10302-10309. DOI: 10.15680/IJIRSET.2015.0506148

[37] Chamundeeswari J. Experimental study on partial replacement of cement by bentonite in paverblock. International Journal of Engineering Trends and Technology. 2012;**3**(6):41-47

[38] Selvaraj R, Priyanka R. Characteristics study of bentonite mortar with partial replacement of cement. International Journal of Engineering Science Invention Research & Development. 2015;**II**:178

[39] Chandrakanth M, Poorna NS, Rao C, Srinivasa Rao K. Experimental studies on concrete with bentonite as mineral admixture. GRD Journal For Engineering. 2016;**1**(2):7-11

[40] Aravindhraj M, Sapna BT. Influence of bentonite in strength and durability of high performance concrete. International Research Journal of Engineering and Technology. 2016;**3**(5):3120-3125

[41] Ahad MZ, Ashraf M, Kumar R, Ullah M. Thermal, physico-chemical, and mechanical behaviour of mass concrete with hybrid blends of bentonite and fly ash. Materials (Basel). 2018;**12**(1). DOI: 10.3390/ma12010060

[42] Syam Ananya Lia Joe AS. Partial Replacement of Fine Aggregate with

Demolished Concrete Fine Aggregate and Partial Replacement of Cement with Bentonite. 2016. Available from: www.ijste.org

[43] Ghonaim S, Morsy R. Study of bentonite usage in environmentally friendly concrete. Journal of Al-Azhar University Engineering Sector. 2020;**15**(57):1012-1024. DOI: 10.21608/auej.2020.120366

[44] Ahmad J et al. A step towards sustainable self-compacting concrete by using partial substitution of wheat straw ash and bentonite clay instead of cement. Sustainability. 2021;**13**(2):1-17. DOI: 10.3390/su13020824

[45] Taklymi SMQ, Rezaifar O, Gholhaki M. Investigating the properties of bentonite and kaolin modified concrete as a partial substitute to cement. SN Applied Science. 2020;**2**(12):1-14. DOI: 10.1007/s42452-020-03380-z

[46] Xie Y, Li J, Lu Z, Jiang J, Niu Y. Effects of bentonite slurry on air-void structure and properties of foamed concrete. Construction and Building Materials. 2018;**179**:207-219. DOI: 10.1016/j.conbuildmat.2018.05.226

[47] Karthikeyan M, Raja Ramachandran P, Nandhini A, Vinodha R. Application on partial substitute of cement by bentonite in concrete. International Journal of ChemTech Research. 2015;**8**(11): 384-388

[48] Neville AM. Properties of Concrete. 2009

[49] Sreeniva SK, Praneeth SKP, AchyuthA KRM, Ranga RV. A study on durability of concrete by partial replacement of cement with bentonite. International Journal of Chem-Tech Research. 2017;**10**(6):898-904

[50] Swarup OK, Reddy PVR, Achyutha M, Reddy K, Rao VR. A study on durability of concrete by partial replacement of cement with bentonite and fly ash. International Journal of Chemtech Research. 2017;**10**(7):855-861

Chapter 3

The Effects of Mill Conditions on Breakage Parameters of Quartz Sand in the District of Şile on the Black Sea Coast of İstanbul

Serhan Haner

Abstract

Casting, glass, ceramic, construction, plastic, dyeing, and abrasive industries are the main consumption areas of quartz sand, which are formed as a result of the weathering of igneous metamorphic rocks. In such industries, it is very important to select the correct ball size in order to grind the raw material to the desired particle size in optimum time. In this study, the changes in the specific rate of breakage of the quartz sand sample were investigated by using alloy steel balls of five different sizes. For this purpose, three different mono-size samples were prepared according to $\sqrt[4]{2}$ series in the range of 0.090–0.053 mm. The quartz sand prepared in these three intervals was ground with 6.35, 7.94, 9.52, 12.70, and 19.05 mm alloy steel balls for different durations. The specific rate of breakage values was obtained from the particle size distributions acquired after various grinding periods. As a result of grinding tests, an increase in the rate of breakage is observed due to the increase in ball diameter.

Keywords: quartz sand, breakage, ball size, comminution, Şile, İstanbul

1. Introduction

Turkey has 2.5% of the global industrial mineral reserves, 73% of the global boron mineral reserves, 20% of the global bentonite reserves, and more than half of the global perlite reserves. The mines extracted from these sources are used as raw materials in the industry, with the excess being exported. Around 791 million tons of industrial minerals are produced worldwide, and Turkey accounts for 42.3 million tons of this global production. Turkey ranks 3rd in the world with a share of 5.3% in industrial mineral production. When we consider this production rate in terms of value, it ranks 8th with a 4 percent share. Based on the figures for 2016, the mines extracted most in Turkey's industrial raw material production were calcite, feldspar, gypsum, quartz sand, pumice, and boron. These production data were drawn up based on the production amount figures declared by licensed mine sites to the Turkey General Directorate of Mining and Petroleum Affairs.

Quartz sand deposits are very common in Turkey. There are quartz sand deposits in İstanbul-Şile and Çatalca, Zonguldak-Kilimli, Bartın, Tekirdağ-Safaalan and Sinop-Sarıkum. In Turkey, there are 1.884.208.585 tons of (visible+probable) quartz sand reserves containing over 90% SiO_2. A total of 54.820.154 tons of quartz sand were produced in Turkey between the years 2011 and 2016.

Quartz sand is formed as a result of the decomposition of quartz-rich magmatic metamorphic rocks. Quartz sand is divided into two types based on its formation. Magma-origin rocks have decomposed and weathered where they formed by physical forces such as the atmosphere and faults. These deposits have higher SiO2 content. Some other deposits piled up in one area during being moved and formed placer beds. During transportation, heavy minerals also collapsed and turned into deposits when moving with the silica. Quartz sand consists of silica particle of 1/16 and 2 mm size. Its pure one is white in color. On the other hand, depending on the amount of iron minerals (limonite, pyrite, magnetite, hematite, etc.) in it, it can be brown, red, or pink in color. It contains a high amount of silica. Although it can be found pure in nature, it may contain small amounts of clay, feldspar, iron oxides, or carbonates. The beneficiation processes such as gravity, frotation, and leaching are applied in order to bring the requested chemical, physical, or thermal properties depending on the intended use. According to their intended use, quartz sands are generally named core sand, glass sand, golf course sand, hydraulic fracturing sand and blasting sand. In determining the usage area of quartz sand, it is important to know the maximum chemical impurity and minimum SiO_2 levels, and the features such as particle size distribution and grain shape. There must be at least 95% SiO_2 in quartz sand to be used in the production of casting mold, silica bricks, silicone, ferrosilicon, and building sand, and there are certain limit levels for Al_2O_3 and Fe_2O_3 content. Quartz sand is in a general sense used in the glass and casting industry. Apart from these areas of application, it is also used in industries such as construction, aerated concrete, ceramic, iron-steel, dyeing, plastic, and abrasive, which is used for removing rusted surfaces, corroded surfaces, old paint, as well as for shaping marble and glass. The open-pit mining method is applied as the production method from the pit. For quartz sand production to be economical, the ratio of the thickness of the cover layer to the thickness of the quartz sand layer should not exceed the 4 m^3/ton level [1–4].

In this study, quartz sand in the district of Şile on the Black Sea Coast of İstanbul, Turkey, which is used in the production of traditional ceramic materials, is preferred. Over 4 million tons of quartz sand are produced annually from the Şile Basin and utilized in many fields in Turkey. Şile region quartz sand reserves are estimated to be over 100 million tons. In the traditional ceramics industry, quartz sand containing 90% > SiO_2 and Fe_2O_3 < 0.5% with a particle size of approximately −1 + 0.075 mm is preferred. The quartz sands of the Şile region generally have the characteristics to meet these expected oxide properties. Mining companies sell these quartz sands only after they have been washed and classified. Non-plastic raw materials such as feldspar and quartz sand supplied by traditional ceramic factories are also applied to the grinding process. Generally, an alumina ball is used as the grinding medium in the grinding process to give ceramic materials the desired physical, chemical, thermal and mechanical properties. Ball mills are preferred for intermediate grinding (P80; 0.040 to 0.40 mm) in plants producing traditional ceramic products such as tile, sanitaryware, tableware, and. The non-plastic composition is ground to a particle size finer than 0.075 mm with the grinding process. After the grinding process, the Fe_2O_3 content in the ceramic sludge is removed with magnetic holders.

Upon examination of the quartz sand of the Şile region, which was supplied for use in the studies, with a loop, dark-colored ferrous minerals with a grain size of approximately 0.010–0.040 mm were discovered. These minerals, which do not pose a significant problem in the traditional ceramic industry, can be considered as an essential impurity in areas such as glass and casting mold production. If it is necessary to obtain quartz sands with higher SiO_2 and lower Fe_2O_3 content, such as glass and casting mold production, enrichment processes must be applied. In such cases, quartz sands are enriched by gravity method, flotation, or extraction according to their intended use, and the impurities it contains are removed. In the gravity method, first of all, the clay minerals that form slime must be cleaned. Then, minerals with magnetic properties such as hematite, magnetite, or ilmenite must be removed with magnetic separators of 1000–15,000 gauss intensity. Wet magnetic separators are preferred for cleaning magnetic minerals smaller than 0.075 mm. Enrichment by gravity becomes increasingly challenging as the grain size decreases. As in the quartz sand of the Şile region used in this study, some impurities can be liberated in grain sizes below 0.075 mm. In such cases, the quartz sand must first be ground into the liberation size. Afterward, enrichment by flotation is required [3, 5]. Since the grinding process is under a specific particle size, most of the energy used is converted into heat energy. While specific energies of ball mills increase exponentially in these fine particle sizes, the grinding efficiency decreases economically [6–10].

Quartz sand acts as a grinding medium on other non-plastic raw materials that form the ball mill phase in ceramic materials production. Raw materials such as feldspar in the mill show fracture along smooth surfaces because they have cleavage. However, since quartz sand is no cleavage, it does not show the smooth fracture. Fracture occurring along irregularly developed cracks in quartz sand takes place conchoidally (mussel shell) [11]. As a result, when compared to other raw materials in the mill, quartz sand grinding and the energy consumed during this operation are significantly high. It also causes some wear on grinding media such as quartz sand, alumina ball, and flint pebbles. In this study, alloy steel balls were favored as a grinding medium in grinding units above alumina balls and flint pebbles, which are preferred by ceramic producers. Alloy steel balls have approximately twice the specific gravity of alumina balls. The grinding medium's weight applied to the unit volume during the grinding process and the size of the grinding medium are critical elements determining the mills' capacity and efficiency. There are some studies in the literature on the selection of the grinding medium size in the ball mill [6, 12–17]. In addition, there are studies of breakage rate parameters of some raw materials [18–20].

In this study, the effect of different sizes of alloy steel balls on specific rates of breakage (S_i) was investigated. The quartz sand used in grinding works was supplied by a private mining company, which is located in the district of Şile, Istanbul. The variation of the specific rate of breakage of quartz sand was investigated using 6.35, 7.94, 9.52, 12.70, and 19.05 mm alloy steel balls. For the grinding tests carried out in dry conditions, the powder (f_c = 0.120) and ball loads (J = 0.35) are taken as fixed. For this purpose, the kinetic model, the basis of which was developed by Austin et al. (1984), was applied. In this model, mathematical expressions are defining the breakage distribution and breakage rate of raw material [21]. Studies with kinetic model-based grinding and the values obtained in the laboratory are suitable for simulation in an industrial environment [22].

2. Materials and methods

Quartz sand used in laboratory-scale grinding experiments was obtained from a private mining company located in Şile, on the Black Sea coast of İstanbul. Chemical analysis values of quartz sand are given in **Table 1**.

In this study, specific rate of breakage values of Şile region quartz sand in three different mono-size intervals were determined. For this purpose, Şile region quartz sand was prepared in mono-size intervals of −0.090 + 0.075, −0.075 + 0.063, −0.063 + 0.053 mm according to the $^4\sqrt{2}$ sieve series. In order to determine the specific rate of breakage values of quartz sand, a 150x150 mm (diameter x length) stainless steel ball mill was used as the grinding medium. The diameters of the grinding balls in this grinding medium were chosen as 6.35, 7.94, 9.52, 12.70, and 19.05 mm. In order to determine the specific rate of breakage values of quartz sand in three mono-size intervals, it was ground in batches at certain time periods (1, 2, 4, 8, 16, 32, and 64 minutes). After each grinding period, all the powder in the mill was discharged, and representative samples were taken. A laser diffraction device was used to measure the particle sizes of the representative samples belonging to the grinding periods. Based on each time period of grinding, semi-logarithmic graphs of the material fractions staying in the high points of the particles' size limits were drawn in contact with the grinding periods. The first-order zone breakage is represented by the zone in which this graph decreases linearly. The slope of the line in the first-order breakage zone gives us the specific rate of breakage of the material in that particle size range. The formula for the specific rate of breakage (S_i) is given in Eq. 1.

$$S_i = a(\mathrm{x_i}/1\ \mathrm{mm})^{\alpha} Q_i \tag{1}$$

The symbol "a" given in Eq. 1 is the model parameter. This parameter depends on the mill conditions. "α" value is a positive number, normally in the range 0.5 to 1.5, which is characteristic of the material properties. "x_i" symbolizes the upper dimension (mm) in the fraction i. Q_i is the correction factor and is taken as 1 for smaller sizes. The experimentally established values of Q_i can be seen in Eq. 2 [21].

$$Q_i = \frac{1}{1 + (x_i/\mu)^{\Lambda}} \qquad \Lambda \geq 0 \tag{2}$$

SiO_2	91.16
Al_2O_3	5.18
Na_2O	0.62
K_2O	0.37
CaO	0.05
Fe_2O_3	0.34
TiO_2	0.43
SO_3	0.03
Loss on ignition	1.82

Table 1.
Chemical composition of quartz sand, mass-%.

Eq. 2 refers to the fact that "μ is the particle size at which the correction factor 1/2 and Λ a positive number which is an index of how rapidly the rates of breakage fall as size increases (the higher the value of Λ, the more rapidly the values decrease)" [21].

In laboratory grinding studies, the rotational speed of the ball mill was chosen to be 70% of the critical speed value of the ball mill. The critical speed of the ball mill was calculated using Eq. 3. Amounts of ball and material to be fed to the mill with Eqs. 4–6 respectively and the mill's interstitial filling rates were found.

$$\text{Critical speed } (N_c) = \frac{42.3}{\sqrt{(\mathrm{D} - \mathrm{d})}} \tag{3}$$

In Eq. 2, D represents the internal mill diameter (m) and d represents the maximum ball diameter (m) [21].

$$J = \frac{\text{Mass of balls/Ball density}}{\text{Mill volume}} * \left(\frac{1}{0.6}\right) \tag{4}$$

$$f_c = \frac{\text{Mass of powder/Powder density}}{\text{Mill volume}} * \left(\frac{1}{0.6}\right) \tag{5}$$

$$U = \frac{f_c}{0.4 * J} \tag{6}$$

The properties of the ball mill, experimental conditions, alloy steel balls, and quartz sand used in the laboratory grinding process are given in **Table 2**.

Shoji et al. (1982) found a simple relationship between powder filling and ball load in the mill [23]. It is seen in Eq. 5. In Eq. 6, the net mill power (m_p) as a function of ball load was fitted by the empirical function. Eq. 7, which is the combination of Eqs. 5 and 6, was used to calculate the specific grinding energy as a function of ball filling. Combining Eqs. 5 and 6 gives the result shown in **Figure 1** [21].

$$S(f_c, J) \propto a \propto \frac{1}{1 + 6.6J^{2.3}} \exp\left[-cU\right], 0.5 \le U \le 1.5, 0.2 \le J \le 0.6 \tag{7}$$

Mill	Diameter, D mm	150
	Length, mm	150
	Volume, mm^3	2650x10^3
Mill speed	Critical (N_c), rpm	111–117
	Operational ($Ø_c$ = 70%), rpm	78–82
Ball	Quality	Alloyed steel
	Specific gravity, g/cm^3	8.09
	Diameter, d mm	6.35, 7.94, 9.52, 12.70, 19.05
	Fractional ball filling, J	0.35
Material	Specific gravity, g/cm^3	2.65
	Fractional powder filling, f_c	0.12
	Powder-ball loading ratio, U	0.86

Table 2.
Ball mill characteristics and test conditions for grinding of quartz sand.

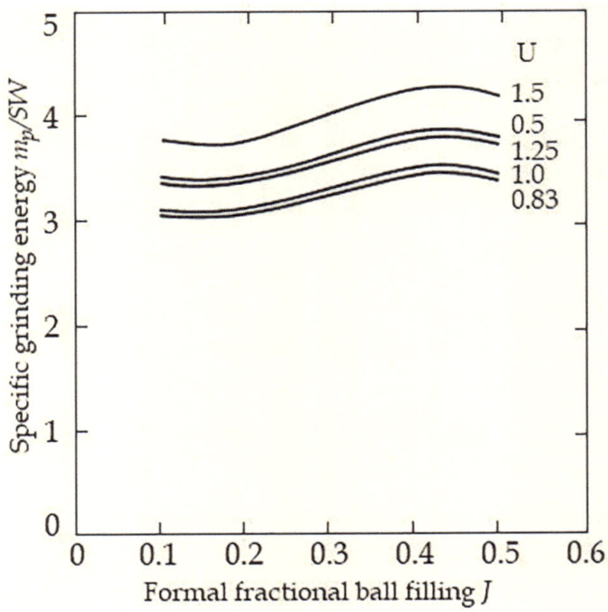

Figure 1.
Relative specific grinding energy as a function of ball filling: dry grinding in a laboratory mill [18].

where c is 1.32 and 1.2 for wet and dry grinding respectively.

$$m_p \propto \frac{1-0.937J}{1+5.95J^5}, \quad 0.2 \le J \le 0.6 \tag{8}$$

$$\text{Specific grinding energy} \propto \left\{\frac{1-0.9375J}{1+5.95J^5}\right\} / \left\{\frac{Ue^{-1.2U}}{1+6.6J^{2.3}}\right\} \tag{9}$$

Austin et al. (1984) express the connection between specific grinding energy and ball load as follows: "Although the capacity of a laboratory mill is a maximum at 40 to 45% ball load, the relative specific grinding energy m_p/SW is a minimum at about 15 to 20% ball load. In practice, ball loads less than 25% are not normally used because low ball loads can give excessive liner wear. In addition, mill capacity is clearly lower for lower ball loads [21]." Dependent upon Eq. 7 and **Figure 1**, the values of specific grinding energy changes based on the ball loads J and powder-ball loading ratio U. As a result, in this study the specific grinding energy values are obtained from Eq. 7 for 0.35 ball filling ratio was calculated as 3.38.

3. Results and discussion

The quartz sand in the three different mono-size intervals were grinding linearly with increasing grinding times. At the end of each milling period, the fractions of material remaining in the top particle size range were plotted against milling times. The graphs of the first-order breakage lines obtained for five different ball sizes are given in **Figures 2–4**. The region where the graph decreases linearly represents the first-order breakage region. The slope of the line in the first-order breakage region gives the specific rate of breakage based on the particle size range of the quartz sand.

After determining the specific rate of breakage for the three mono-size intervals fractions exhibiting first-order breakage kinetics behavior, S_i values were plotted against particle size fraction. The rate of breakage parameters of these lines was determined as a_T and α. The results are given in **Figure 5**.

a_T and α values, which are parameters of specific rates of breakage, were obtained by non-linear regression (from Eq. (1) and **Figure 5**), and are 0.15, 0.85 for 6.35 mm and 0.14, 0.78 for 7.94 mm and 0.14, 0.76 for 9.52 mm and 0.24, 0.95 for 12.70 mm and 0.26, 0.96 for 19.05 mm, respectively. In **Figure 5**, specific breakage rates also decreased based on the decrease in alloy steel ball size in general. Moreover, when the graphics in **Figure 5** are evaluated based on the particle size, the breakage rates

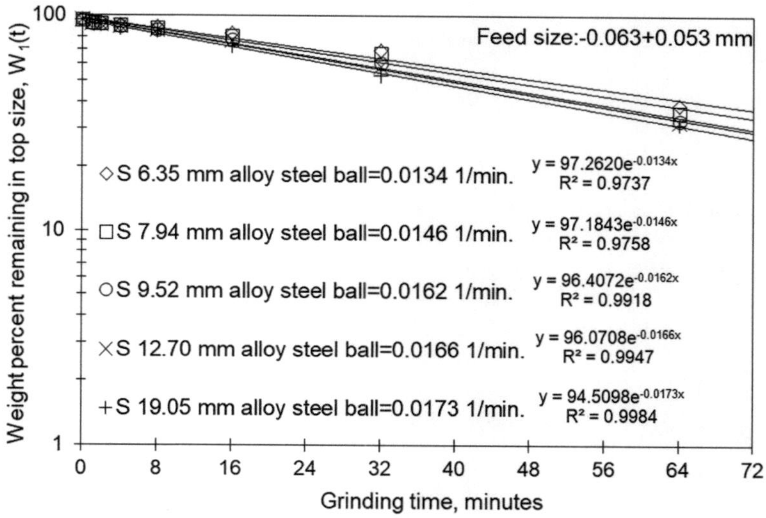

Figure 2.
First-order plots for alloy steel balls with different diameters of quartz sand as well as −0.063 + 0.053.

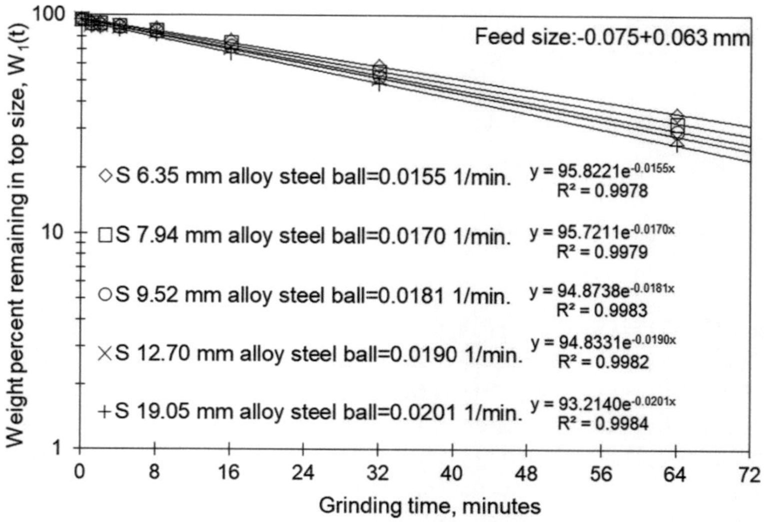

Figure 3.
First-order plots for alloy steel balls with different diameters of quartz sand as well as −0.075 + 0.063.

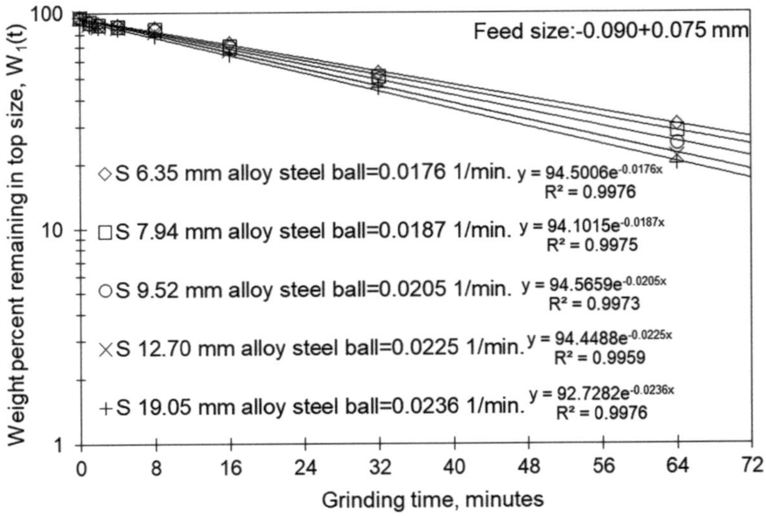

Figure 4.
First-order plots for alloy steel balls with different diameters of quartz sand as well as −0.090 + 0.075.

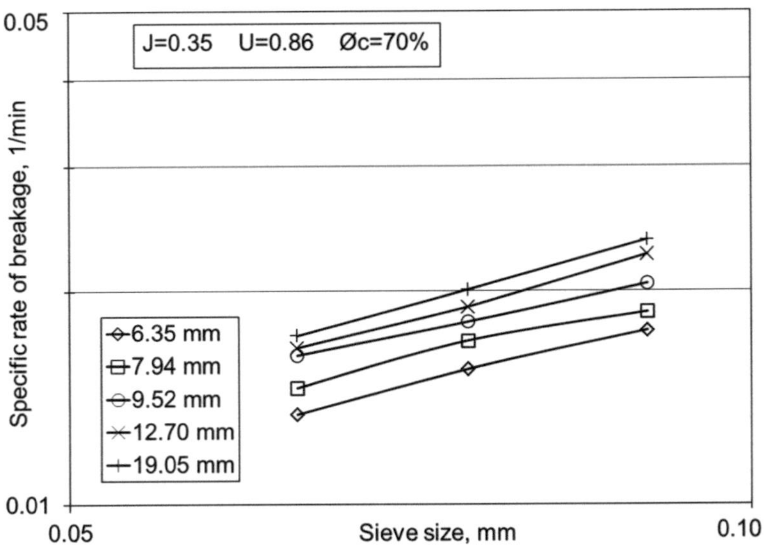

Figure 5.
Variation of the specific rate of breakage as a function of the maximum feed size for quartz sand ground with different alloy steel balls.

decrease as the particle size intervals decrease. The presence of a maximum is quite logical because large lumps obviously will be too strong to be broken in the mill. Austin et al. (1984) explain that "[t]he theory of fracture implies that smaller particles are relatively stronger because larger Griffith flaws exist in larger particles and they are broken out as size is reduced. The fact that the specific rates of breakage are a simple power function of size has not been adequately explained on a theoretical basis, but it has been amply demonstrated by many experiments [21]."

In ball mills, large balls are known to be responsible for the breakage of coarse particles, and small balls are supposed to grind the fine ones. Austin et al. (1984)

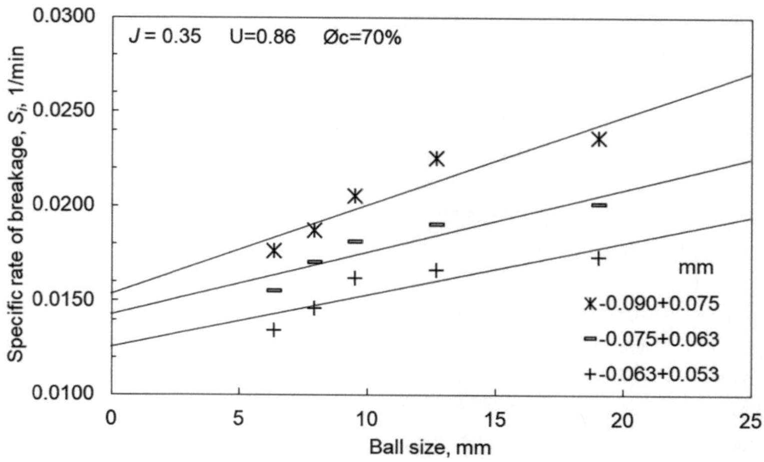

Figure 6.
Variation of ball diameter with first order breakage constant.

stated the effect of ball diameter on breakage rate as "considering a representative unit volume of the mill, the rate of ball-on-ball contacts per unit time will increase as ball diameter decreases since the number of balls in the mill increases as $1/d^3$. Thus, the rates of breakage of smaller sizes are higher for smaller ball diameters [21]." However, balls in the range of 20 mm–50 mm were used for grinding the raw material with a particle size of 30x40 mesh here. The grinding conditions in this study are not the same. According to $\sqrt[4]{2}$ sieve series, 3 mono-sized fractions in the range of −0.090 + 0.053 were used. The grinding process was performed with alloy steel balls in the range of 6.35–19.05 mm. It is understood from the results that the grinding was difficult due to working in too small particle size ranges. In **Figure 6**, it can be said that the grinding energy that large-sized balls transferred on the quartz sand particles is high and therefore, high specific rates of breakage values were obtained in the grinding works carried out with large-sized balls. It can be seen that the grinding process is carried out faster by using large-sized balls compared to small-sized balls. This study was shown that d = 19.05 mm was the optimum balls size for the maximum breakage rates.

4. Conclusions

Quartz and quartz sand consumption in the traditional ceramics industry in Turkey is approximately 600.000 tons per year. The quartz sand, which has the features that can meet the needs of Turkey's ceramics industry in terms of cost and chemical content, is produced in Şile, İstanbul. Glass quality quartz sand beds have decreased around İstanbul. The quartz sand beds in Turkey are suitable for casting and ceramics industries except for glass. Quartz sands separated by washing in some clay deposits are being evaluated. In addition, as a result of the evaluation of side products in the ore dressing facilities in some raw material quarries that are not economically operable, the operation of these quarries will be possible. It is essential to keep the energy consumed in downscaling processes at an optimum level in order to ensure the economy in raw material production. When the literature is examined, it is seen that many raw materials used for different purposes do not have breakage values under different milling conditions. Breakage values of raw materials belonging to any region differentiate according to their properties such as mineral rates in the raw material, its structural features, chemical impurities, and physical fractures. Therefore, in order for the ore dressing facilities to keep the energy consumed in grinding processes at an optimum level, the grinding kinetics of the raw material must be taken into account.

Accordingly, in this study, the impact of ball size on the grinding characteristic of quartz sand in the district of Şile on the Black Sea Coast of İstanbul in the laboratory ball mill was examined. These quartz sands are rather used in the production of traditional ceramics materials. However, if it is enriched by flotation, it can also be used in different industries. Grinding and classification processes must definitely be applied in order for the quartz sands to be used in any other area of the industry. That is, it is desired that the particle size is in a certain range. The grinding process in the traditional ceramic industry is carried out with ball mills and generally uses alumina balls. A large part of the energy consumed in grinding processes carried out in rod and ball mills turns into sound and heat energy. The grinding efficiency decreases due to this situation. It is very important to choose balls with the appropriate size to reduce the inefficiency of grinding.

In this study, it was found that very small ball sizes could not play an effective role in grinding quartz sand and that their impact and attrition effect on the particles was low. The energy transferred by the steel balls to the quartz sand particles during grinding increased with the increase in ball size. In this study carried out with different ball sizes, it was found that the most effective breakage was achieved with d = 19.05 mm alloy steel ball.

Author details

Serhan Haner
Dinar Uygulamalı Bilimler Yüksekokulu, Afyon Kocatepe University, Afyonkarahisar, Turkey

*Address all correspondence to: serhan.haner@gmail.com

References

[1] Ciullo PA. Industrial Minerals and Their Uses. New Jersey: Noyes Publications; 1996. p. 632. No:96-29173

[2] Barhana S, Kılavuz O. Devlet Planlama Teşkilatı TC. Madencilik özel ihtisas komisyonu raporu. Vol. 624. Ankara: endüstriyel hammaddeler alt komisyonu toprak sanayii hammaddeleri III; 2001. p. 2613

[3] Yıldız N. Cevher Hazırlama ve Zenginleştirme. Ankara: Ertem Basım Yayın Dağıtım; 2014. p. 1504. 978-975-96779-5-4

[4] Kalkınma Bakanlığı TC. On birinci kalkınma planı (2019–2023). Vol. 822. Ankara: Madencilik politikaları, Özel ihtisas komisyonu raporu; 2018. p. 3041

[5] SERHAM. 2015–2017 dönemi faaliyet raporu. Seramik, cam ve çimento hammaddeleri üreticileri derneği, İstanbul; 2017

[6] Haner S. The effects of ball size on the determination of breakage parameters of nepheline syenite. Journal of Mining Science. 2020;**56**:848-856. DOI: 10.1134/S1062739120057191

[7] Lidell KS. Machines for fine milling to improve the recovery of gold from calcines and pyrite. Johannesburg: Proceeding of the International Conference on Gold; 1986. pp. 405-417

[8] Bouchard J, LeBlanc G, Levesque M, Radziszewski P, Georges-Filteau D. Breaking down energy consumption in industrial grinding mills. In: Canadian Institute of Mining, Metallurgy and Petroleum. Proceedings 49th Annual Canadian Mineral Processors Operators Conference, Canada; 2017. pp. 25-35

[9] Fuerstenau DW, Abouzeid A-ZM. The energy efficiency of ball milling in comminution. International Journal of Mineral Processing. 2002;**67**:161-185. DOI: 10.1016/S0301-7516(02)00039-X

[10] Schellinger AK. A Calorimetric method for studying grinding in a tumbling medium. Transactions of AIME. 1951;**190**:518-522

[11] Vardar M, Bozkurtoğlu E. Yerkabuğunu Oluşturan Maddeler Mineraller ve Kayaçlar. İnşaat Jeolojisi, 2009–2010 Course Year Grades, İstanbul; 2009. p. 20

[12] Austin LG, Shoji K, Luckie PT. The tffect of ball size on mill performance. Powder Technology. 1976;**14**:71-79. DOI: 10.1016/0032-5910(76)80009-5

[13] Bond FC. Grinding ball size selection. Transactions of AIME. 1958; **211**:592-595

[14] Yusupov T S, Kirillova E A, Denisov G A. Dressing of quartz-feldspar ores on the basis of selective grinding and mechanical activation. Journal of Mining Science. 2003;39: 174-177. DOI: 10.1023/B:JOMI .0000008464.34124.1d

[15] Coghill WH, Devaney FD. Ball Mill Grinding. Washington: United States Government Printing Office; 1937. Available from: https://play.google.com/books/reader?id=k4MbYBy8674C&hl=tr&pg=GBS.PP1 [Accessed: December 15, 2019]

[16] Wolosiewicz-Glab M, Foszcz D, Gawenda T, Ogonowski S. Design of an electromagnetic mill. Its technological and control system structures for dry milling. Poland: E3S Web of Conferences: Mineral Engineering Conference (MEC 2016); 2016. pp. 1-9

[17] Haner S. Fine dry grinding with cylpebs of quartz sand in Şile, district of İstanbul, Turkey. Archives of Mining Sciences. 2021;**66**:624-634. DOI: 10.24425/ams.2021.137457

[18] Barani K, Balochi H. First-order and second-order breakage rate of coarse particles in ball mill grinding. Physicochemical Problems of Mineral Processing. 2016;**52**:268-278. DOI: 10.5277/ppmp160123

[19] Barani K, Balochi H. A comparative study on the effect of using conventional and high pressure grinding rolls crushing on the ball mill grinding kinetics of an iron ore. Physicochemical Problems of Mineral Processing. 2016;**52**:920-931. DOI: 10.5277/ppmp160231

[20] Olejnik TP. Grinding kinetics of granite considering morphology and physical properties of grains. Physicochemical Problems of Mineral Processing. 2012;**48**:149-158

[21] Austin LG, Klimpel RR, Luckie PT. Process Engineering of Size Reduction: Ball Milling. New York, United States of America: American Institute of Mining Metallurgical and Petroleum Engineers Inc.; 1984

[22] Austin LG, Bagga R, Çelik M. Breakage properties of some materials in a laboratory ball mill. Powder Technology. 1981;**28**:235-241. DOI: 10.1016/0032-5910(81)87049-0

[23] Shoji K, Austin LG, Smaila F, Brame K, Luckie PT. Further studies of ball and powder filling effects in ball milling. Powder Technology. 1982;**31**: 121-126. DOI: 10.1016/0032-5910(82)80013-2

Chapter 4

Thermal Conductivity and Mechanical Properties of Organo-Clay-Wood Fiber in Cement-Based Mortar

Fadhel Aloulou and Habib Sammouda

Abstract

This paper orientated to study the compressive resistance and thermal conductivity of compressed and stabilized clay blocks in the cement matrix. The effect of the content of wood fiber (WF) became studied as a reinforcement material in cement mortars. The porosity, compressive energy, thermal conductivity and composite of cement hydration had been investigated. The addition of NFC suggests a very good pore reduction, and the fine result becomes acquired with the emulsion of a combination incorporating 2%wt of WF inside the presence of an anionic surfactant (SDBS). The results revealed that used in this study were a mix of water with ordinary portland cement and organo-clay (OC) modified with Cetyltrimethylammonium bromide at water-to-solid ratios 1%. The effect depending on w/s ratio of OC used samples with cement substitution for organoclay showed from 2% higher compressive strength results than that of the plain cement paste and a decrease of the thermal conductivity by addition of 2%wt of WF from 2.26 to 0.8 W/m °C. It was also observed that with increasing w/s ratio higher amount of cement can be replaced by OC. These analyses have revealed that the presence of WF promoted the hydration, by producing more portlandite and calcium silicate gel.

Keywords: cement, wood fiber, organo-clay, properties, thermal conductivity

1. Introduction

Tunisia has plenty of agricultural waste products such as Alfa fiber, *Posidonia oceanica*, wood and palm fiber. Such as these fibers with many advantages: renewable, cheaper, abundance and easy to use [1] in construction and building materials. These natural fibers are cellular solids with enormous specific surface area, high porosity and low thickness [2, 3]. Currently, various composite based on clay minerals and natural fibers are being tested as cementitious composites in a wide range of applications such as construction. Especially, silica materials have good thermal insulation properties. However, low mechanical [4] resistance. This is the reason that different procedures were created to improve physico-chemical properties of the wood fibers with clay and cement [5–7]. These days, silica aerogels have numerous applications including thermal and / or acoustic insulation, and additionally acoustic protection, natural remediation, aviation and biomedicine [8–12].

Lately, as a result of energy consumption and inside the reason of energy-saving, recently works makes a specialty of developing building materials having alluring mechanical and thermal insulating homes. However, improving the mechanical residences induces a lack of thermal insulation overall performance. Today, area heating and cooling comprise the most significant level of general power fed on in homes [13].

The thermal performance of the building envelope is a key factor in determining the measure of the energy required for the comfort of nature. In this regard, a few examinations consider reason that energy consumption can be improved by joining thermal insulation materials in to walls and roofs. Thermal insulation can be characterized as a material that eases back that slows the heat flow in side rout side the building. For its determination, thermal conductivity is the fundamental property to look for. Currently, the research of bio based materials, of natural fiber, of organoclay were used to reinforce cement, to improve the qualities of cementitious materials, to accelerate the formation and the precipitation of hydration [14–17], works to enhance the mechanical properties of the composite and their resistance. In this sense, numerous scientists investigated clay, which belongs to the phyllosilicate family, was due to the permeable media, in exchange, it could also be mixed with many fibers considered as a material with a wide range of applications. The example of clay-based materials was selected.

In which case, there is good reason to use modified clay such as chlorite, Kaolinite and Illite, are considered as filler in the composite material [18], to enhance the properties of cement thanks to its low cost, availability and its excellent characteristics. Besides, in many studies, it was remarkable to note that the role of organoclay has gained many interests on both academic and industrial research [19] due to the excellent performance as well as the way to modify, to as clay-based composite [20].

It's important to mention that natural fibers has grown develop environmental-friendly construction materials. These fibers are biodegradable, lighter and abundant resources for examples: sisal, Flax, Hemp, Bamboo, coir and others [21]. Despite all the advantages of natural fibers, there are some disadvantages, which have confined their applications in the cementitious composites. Initially, for natural fiber, the interfacial bond existing between the fiber and the cement matrix is relatively weak and about the degradation of fibers in a high alkaline of cement adversely affects the mechanical properties and durability of natural fiber reinforced composites [22]. Additionally, for the organoclay, one of the most problem it that increasing the amount of organoclay conducts to reduction of mechanical propreties.

In this work, concrete based composite materials prepared and reinforced with wood fiber and organoclay (OC) particles is used to improve their mechanical quality and to decrease the alkalinity of the matrix through diminishing the $Ca(OH)_2$. Another is to upgrade microstructures qualities of the grid and to add to the bonding existing between the matrix and the wood filaments [23].

In which case, there is a valid justification to utilize the SDBS as surfactant between the platelets of clay treated with CTAB, organoclay can respond with $Ca(OH)_2$ of the concrete hydration products and form structure additional calcium-silica-hydration (CSH) gel. The advantage of the utilization of organoclay is to improve microstructures and mechanical propreties of prepared materials. The wood fibers were characterized via Zeta potential. Fur there more, organoclay in powder form was characterized via structure X-ray diffraction, FTIR and SEM. The organo-clay prepared by treating clay with cetyl tri methyl ammonium bromide (CTAB) on the mechanical. Thermal conductivity and physical properties of wood fibers reinforced cement composites is also studied and investigated.

2. Materials and methods

2.1 Materials

Wood Fiber (WF) and organo-clay (OC) modified with CTAB were used as reinforcements for the cement matrix-composites (**Figure 1a** and **b**).

The clay platelets used in this investigation is a natural Kaolinite treated with CTAB.

The CTAB (Mw 336.39 g mol^{-1}) is a quaternary alkyl ammonium salt is a cationic surfactant, soluble in water and utilized in the preparation of OC. A bout the SDBS (Mw 348.48 g mol^{-1}) is sodium dodecyl benzene sulfonates is an anionic surfactant utilized in the fabrication of composites.

Ordinary Portland cement (OPC NF P 15–301) was used in all mixes of the cement works of Enfidha (**Table 1**).

2.2 Preparation of Organoclay

Clay was prepared to be modified with CTAB, for 3 hours characterized by XRD and SEM in order to determine the amorphous phase of modified clay. To have organophilic clays, you should introduce 10 ml of chlohydric acid [1M]. Then this acid solution is carried in a temperature of 80°C.

And when this temperature became stable, we introduce 10^{-2} moles of CTAB, which we wish to ionize. After three hours of agitation at 80°C, we introduce 10 g of clay. And after three hours of cationic exchange, the clay is filtered and inserted to eliminate the inorganic cations. Finally, the phases of washes are finished and the clay is dried in the steam room at 80°C.

Many works have shown that CTAB surfactant intercalation does not only change the hydrophilic surface characteristics of the clay but also significantly increases the clay interlayer basal spacing.

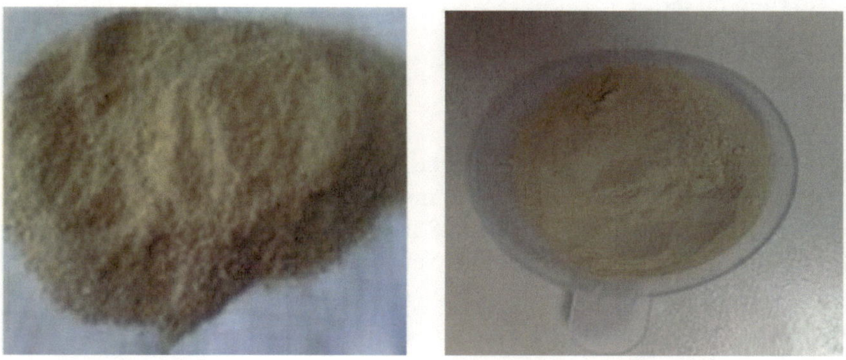

(a) **(b)**

Figure 1.
(a) Wood Fiber (b) Organoclay.

Oxides	Cement (%wt)
CaO	65.47
SiO_2	19.82
Al_2O_3	4.66
Fe_2O_3	3.03
CaO	65.47
MgO	0.84
K_2O	0.64
Na_2O	0.10
TiO_2	0.16
SO_3	2.87
Loss ignition(LOI)	3.5
Density	3.2 g/cm^3
Specific surface area	355 m^2/Kg
Average particle size	18.54 μm

Table 1.
Chemical composition and physical properties of OPC.

2.3 Treatment of wood fibers

To treat the surface of the fiber, the wood fibers were immersed in the aqueous solution of Sodium hydroxide (NaOH) at pH = 12 for 1 hour at 80°C. They were then washed until the pH reached about seven. Finally, the wood fibers were subsequently dried at 80°C for 24 hours. This chemical treatment was aimed at enhancing and improving the mechanical properties by ameliorating the adhesion exciting between fibers/matrix.

2.4 Preparation

2.4.1 Composite materials

The (OPC) was substituted with an organoclay (0.5, 1 and 1.5%wt).

The (OPC) and organoclay were first to dry mixed for 5 min in mixer at low speed and then mixed. The samples were prepared with 0.65 mass ratios a water/organoclay –cement.

2.4.2 Curing and samples

Three series of specimens were prepared of (3.5 × 7 × 1.75) in dimension and were cast in the mechanical tests. All specimens demolded after 24 h of casting and kept underwater for approximately 28 days. Bend test was conducted using an "MTSInsight" Material testing Machine to evaluate, compressive strength.

2.5 Study of physical properties

Porosity and density were measured and conducted to define the quality of composites. The length, thickness, width and weight are measured to determine the bulk density which was carried out by using the following Eq. (1) [20]:

$$D = md/v \quad (1)$$

Where D is the density in g/cm^3,m_d is the mass of the test specimen (g) and V is the volume of the test specimen (cm^3). The porosity Ps was calculated using the following Eq. (2).

$$Ps = (m_s - m_d)/(m_s - m_i) \quad (2)$$

m_i = mass of the sample saturated in water.
m_s = mass of the sample saturated in air.

2.6 Characterization of composite materials

The (OC) prepared was analyzed by X-ray diffraction (XRD) in the 2Θ range between 20 and 80° using CuKα (γ = 1.54060 Å) radiation and was completed by Fourier transform infrared spectroscopy (FTIR) using a Perkin Elmer. The morphology of the organoclay was observed by scanning electron microscopy (SEM) (Hitachi SEM type SU8030 microscope operated at an acceleration voltage of 10 KV and a probe current of 15 pA). The thermal conductivity of different samples was measured by using a "Heat transmission Study Bench – PTC 100".

2.7 Characterization of wood fibers and natural clay

2.7.1 Zeta (ξ) potential

A ξ potential analyzer (Malven 2000) was used to measure the electrophoretic mobility of wood fibers in the aqueous suspension. Measurements were conducted on the fine fraction obtained after filtration of the original fiber suspension through a 45-μm screen. ξ-potential measurement repeated on the whole suspension using the streaming potential technique (Mute kSZP06,using a 40-μm screen as an electrode) matched ξ-potential values obtained by electrophoresis. The ξ-potential values reported are the average of four measurements.

2.7.2 X-ray diffraction (XRD)

The X-ray diffraction patterns were measured with an X-ray diffractometer using CuKα radiation at 40 KV and 30 mA.

2.7.3 Fourier transform infrared spectroscopy (FTIR)

Fourier transform infrared spectroscopy (FTIR) was used to analyze the change of functional groups at the wood fibers and the natural clay after treatment with NaOH, sulfuric acid and CTAB respectively. This test was carried out with a spectrum FTIR, Perkin Elmer. Samples were measured at room temperature and a resolution of 4 cm^{-1}. All FTIR measurements were done in transmittance mode after baseline correction.

2.7.4 Scanning electron microscopy (SEM)

Samples of natural clay and organoclay for Scanning electron microscopy were prepared and analyzed with Scanning Electron Microscopy (Hitachi SEM type

SU8030 microscope operated at an acceleration voltage of 10 KV and a probe current of 15pA).

2.7.5 Thermal conductivity

The measurements of the thermal conductivity characterize the ability of materials to conduct heat energy. The thermal conductivity of different samples of dimensions (24.5 × 1.5 × 24.5 cm) was measured by using a "Heat Transmission Study Bench – PTC 100".

3. Results and discussion

3.1 Particle size of wood fiber

The particles of wood fiber were clearly in the scale range with size 40 μm. The displayed a monomodal size distribution (**Figure 2**) which was relatively narrow (PDI) around 0.2.

3.2 Zeta (ξ) potential

The process of the evolution of (ξ)-potential as a function of pH of the cement particles saturated with the SDBS surfactant are showing in **Figure 3**. The introduction of the SDBS during the preparation of samples passes by several stages. Indeed, for a low concentration of SDBS in order of 0.1 mmol/L, the (ξ) potential is 35 mV. This value results from the ionization of calcium during the hydration of the cement. When the concentration of the SDBS solubilized, Zeta potential decreased. This allows us to deduct the essential role of SDBS in the hydration of the cement, thus in the neutralization of these ions (Ca^{2+}).

Figure 3 shows the quantity adsorbed of surfactant as a function of initial concentration. We notice the existence of three zones:

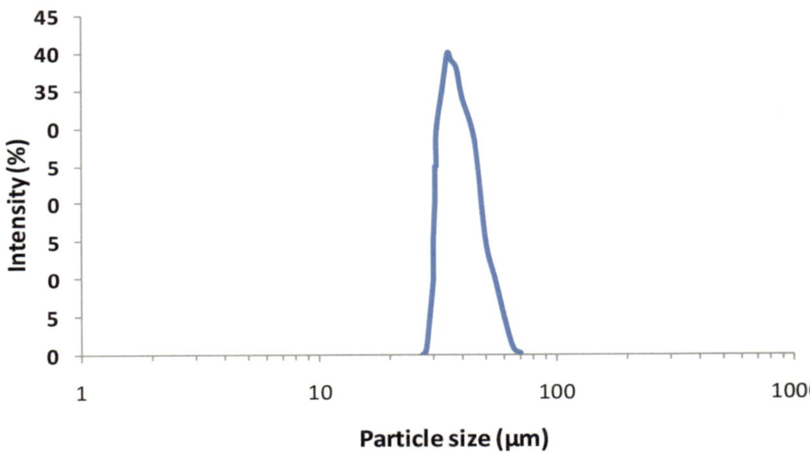

Figure 2.
Particle size of wood fibers.

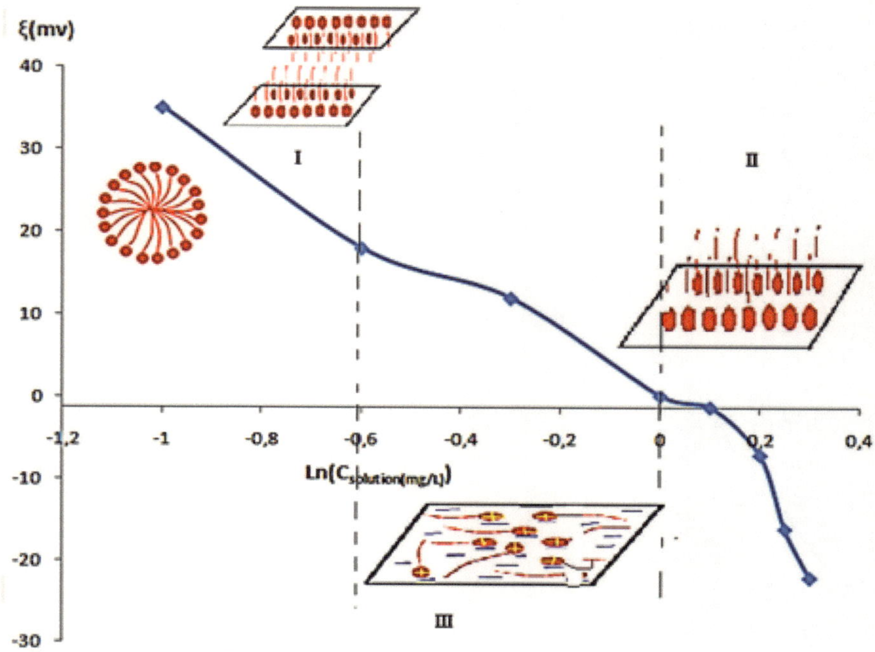

Figure 3.
Variation of the zeta potential according to the initial concentration according to the initial concentration by surfactant.

The first one for low concentration **(I)** we can see equilibrium. This phenomenon was attributed to the micelles, which gives an idea about the formation of micelles beyond a critical concentration. From a certain critical concentration, the hydrophobic interaction between the surfactant molecules of surfactant becomes important compared to the hydrophobic surfactant/water interactions that form spontaneously an association.

The second zone **(II)**, we observe an increase of the slope which is explained by the intercalation of the surfactant molecules with water to form the saturation of the first layer [22]. It is the beginning of mineralization. Thus we observed agglomerations of micelles together, this part is called the micellisation phase.

In the third zone **(III)**, of SDBS, we have (ξ) potential of negative charge (**Figure 4**).

In this case we can say that we have a neutralization of material by the existence of the electrostatic strengths. (The surfactant is cationic and the surface of fibers is negatively charged). The molecules of surfactant are adsorbed at the air/water interface and the superficial concentration increases. From a certain value, a monomolecular layer of surfactant occupies the surface and interfacial tension decreases linearly with the logarithm of the concentration.

3.3 XRD

The XRD patterns of natural clay and organoclay modified by the CTAB are shown in (**Figures 5 and 6**). It can clearly be seen that the natural clay characteristics peaks are present sat 2θ = 7.05°, 12.55°, 25.5° these peaks refer to the presence of Kaolinite

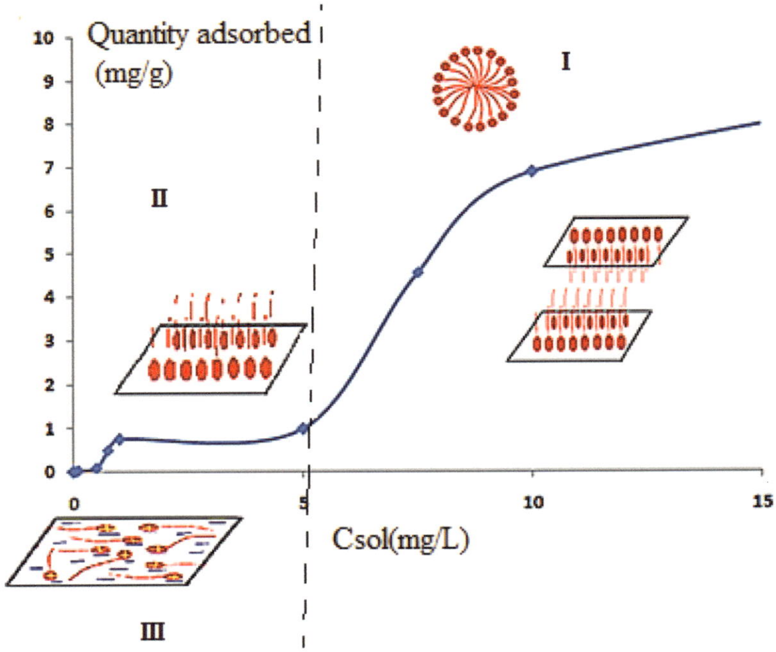

Figure 4.
Evolution of the quantity adsorbed by additive according to the concentration insolution.

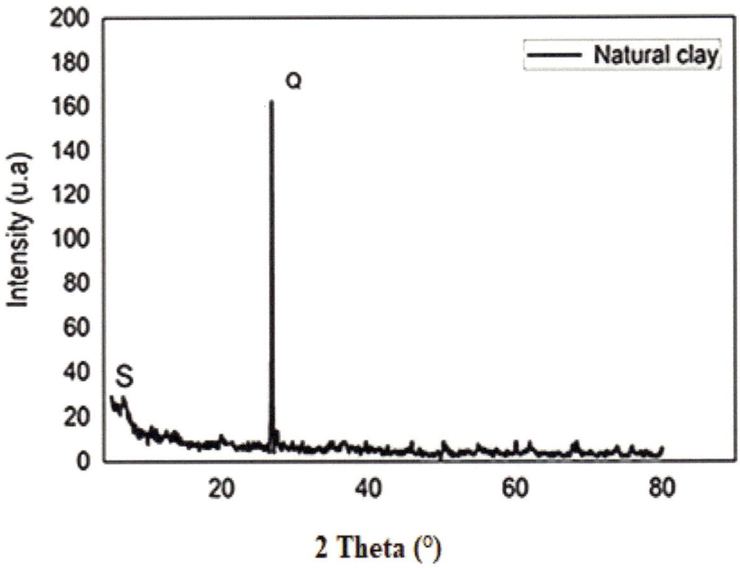

Figure 5.
XRD patterns of natural clay.

the peak at 2θ = 12,55° corresponding at Kaolinite contending Illite. Typical XRD patterns of organoclay after the treatment with the CTAB appears many diffraction peaks at 5.13° until 20°, this result due to the linear structure of CTAB (19 carbons). The CTAB Kaolinite matrix has the highest value of the interfoliar space d001 = 19.10 A° corresponds to the interplanar spacing [23].

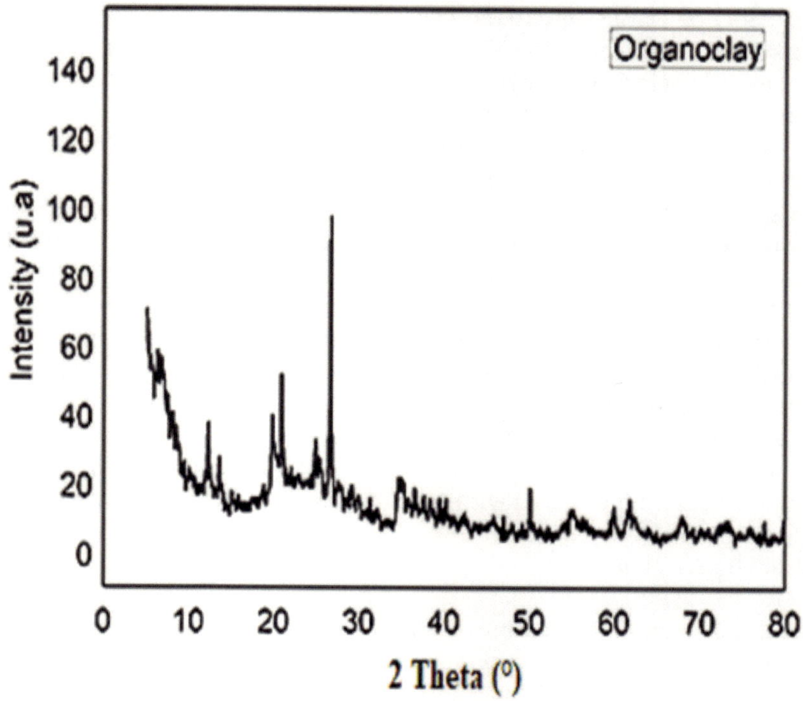

Figure 6.
XRD patterns of Organoclay.

3.4 FTIR of wood Fibers

The FTIR spectra in the range 4000 cm–500 cm^{-1} of the different samples are shown in **Figure 7**. The main characteristic bands of non-treated and treated wood fibers are listed as follows:

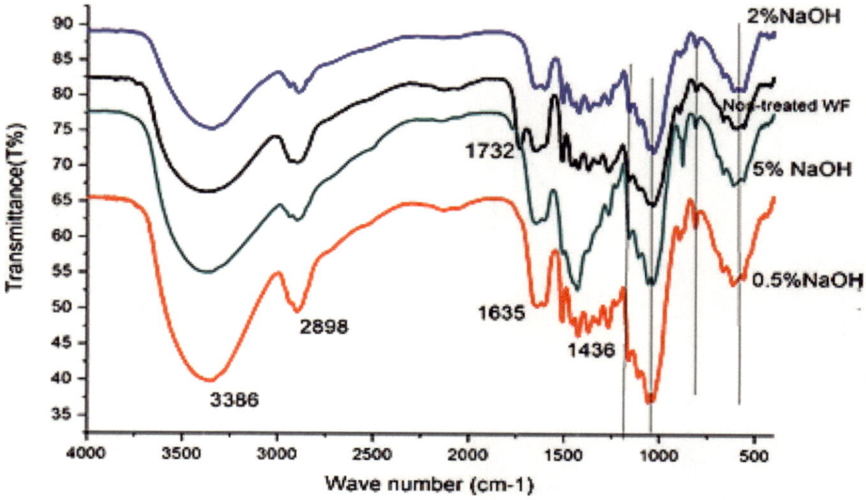

Figure 7.
FTIR spectra of the non-treated wood and the treated wood with NaOH.

The presence of a large band in the 3386 cm^{-1} corresponds to hydroxyl group characteristics of polysaccharides. The band sat 2930 and 2898 cm^{-1} are due successively to sym and usym CH_2 in polysaccharides and fats. The FTIR spectrum exhibits the presence of carbonyl and acetyl groups in the xylan component of (C=O stretching vibration) at 1732 cm^{-1}. However, this peak almost disappears when these fibers are treated with 2% of NaOH. The band at 1635 cm^{-1} is characterized by the vibration of water molecules. Furthermore, the band at 1436 cm^{-1}, assigned to the asymmetric C-H deformation in lignin and hemicelluloses structures. Concerning the FTIR of (WFNT), the peak at 1512 cm^{-1} is indicative of the presence of lignin and is attributed to the C=C aromatic skeletal vibration. In the spectra of WF treated 5% NaOH, this peak was reduced, due to the elimination of lignin by chemical treatments. The small peak sat 1375 cm^{-1} in the spectrum of untreated WF, WF treated 0.5%, 2% are related to CH2 vibration. The band at 1168 cm^{-1}, which appears in all the FTIR spectra, corresponds to the C-O-C asymmetric stretching of the hemicelluloses and lignin. The peak at 1042 cm^{-1}, is assigned to ether linkage (C-O-C) from lignin or hemicelluloses. The peak at 810 cm^{-1} is associated with cellulose, the C-H rock vibrations the cellulose.

3.5 FTIR of Organoclay

Figure 8 shows the characteristics of natural clay and OC treated with CTAB. The peak assignments in the spectra represented OH stretching vibration (3624–3390 cm^{-1}. The treatments of the natural clay with the CTAB make the appearance of two peaks have (2921–2881 cm^{-1}) two peaks attributed to OH stretching vibration. The bond OH (1633 cm^{-1}) can be attributed to water molecules adsorbed on the biomaterial structure.

FTIR exhibits the existence, of a strong band in the range of 750–400 cm^{-1} it was associated with the characteristic Si-O-Si stretching vibration of pure clay. Peaks at 1420 cm^{-1} and 3434 cm^{-1} corresponding to the O-H stretching vibration. Peaks at 1263 cm^{-1}, 2866 cm^{-1} and 2920 cm^{-1} corresponded to the stretching vibrations of –CH, –CH3 and –CH2 respectively. Moreover, characteristic band at 1470 cm^{-1} is assigned to the symmetric vibrations of the COO^- group in the main chain of 1634 cm^{-1} is attributed to the OH bending vibration in the water chemically bond.

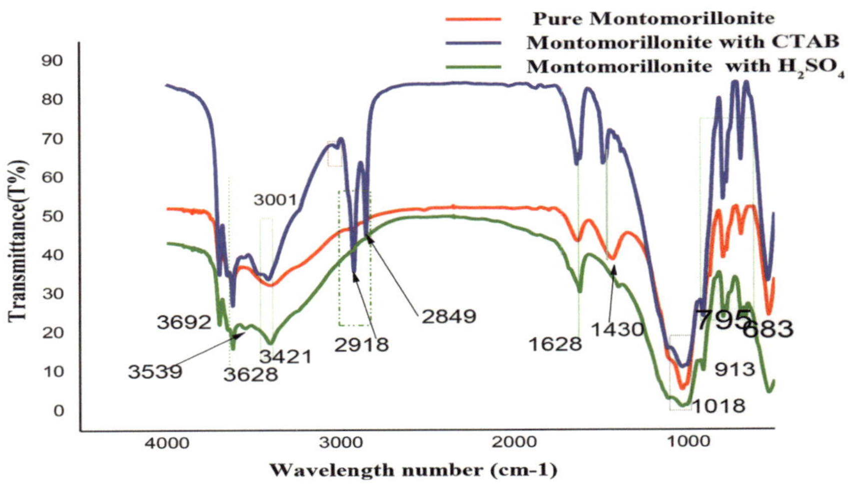

Figure 8.
Infrared spectral characteristics of natural clay and clay treated with CTAB.

3.6 Scanning electron microscopy (SEM)

The micrograph (**Figure 9**) shows the morphology of the surface of raw clay and clay treated with the CTAB, modified by the addition of a low percentage of silica gel. This figure showed that the morphology of the clay surface appeared in the form of plaques, these plaques, of size of 200 nm is plied the some on the others in a characteristic package of structure sheet.

After this treatment by the CTAB and its modification by the addition of a silica gel, we can notice the good separation of some of plaques, layers, also, the appearance of spheres between these leaves in the micrometric size which asserts clearly that the silica gel is well inserted in to the OC.

3.7 Density and porosity

The density values of WF reinforced cement are shown in **Figure 10**. Generally, the composites containing WF exhibit low density than these without WF. This could be attributed to the formation of void sat the interfacial are as between WF and cement matrix. For composite materials with 10% wt of WF, the density decreases by 43%. This indicates that WF has a filling effect on the density of cement composites with or without WF, where the density of cement composites is decreased by the addition of hemp fiber. In **Figure 10** the addition of 10% wt of WF decreased the density of cement composite. That improvement indicated that the addition of WF leads to decrease density and to obtain a composite material with a consolidated microstructure.

The results of porosity and water absorption of values of cement paste, WF reinforced composite, WF–OC reinforced composites are shown in **Figure 11**. Generally, the composites containing WF exhibit higher porosity than that without WF. This could be assigned to the formation of the voids at the interfacial are as between WF and matrices. For composite with 2 and 4% wt of WF, the porosity decreases by 1.82% (4%wt (WF)). The porosity of these composites indicates that WF has a felling effect on the porosity of cement paste composite [24] and this 2% wt and 4% wt are capable of saturating the surface and of reducing pores.

Figure 11 shows that the addition of OC in the composite materials has reduced their porosity. The optimum addition was found as1% wt of OC, which decreased

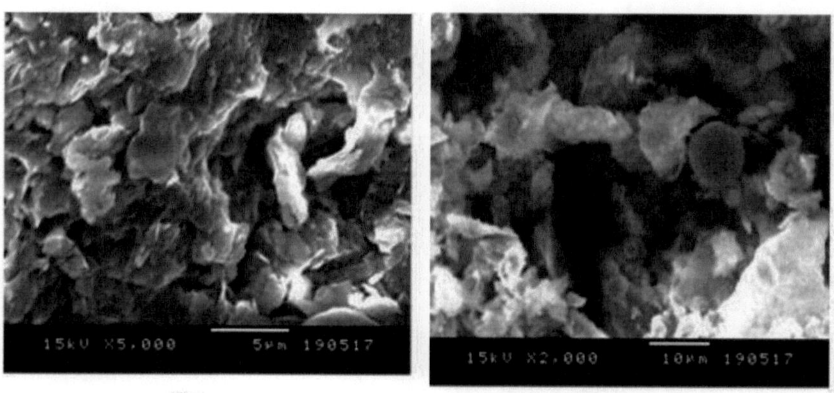

(a) (b)

Figure 9.
MEB micrographs of (a) natural clay and (b) Organoclay with silica gel.

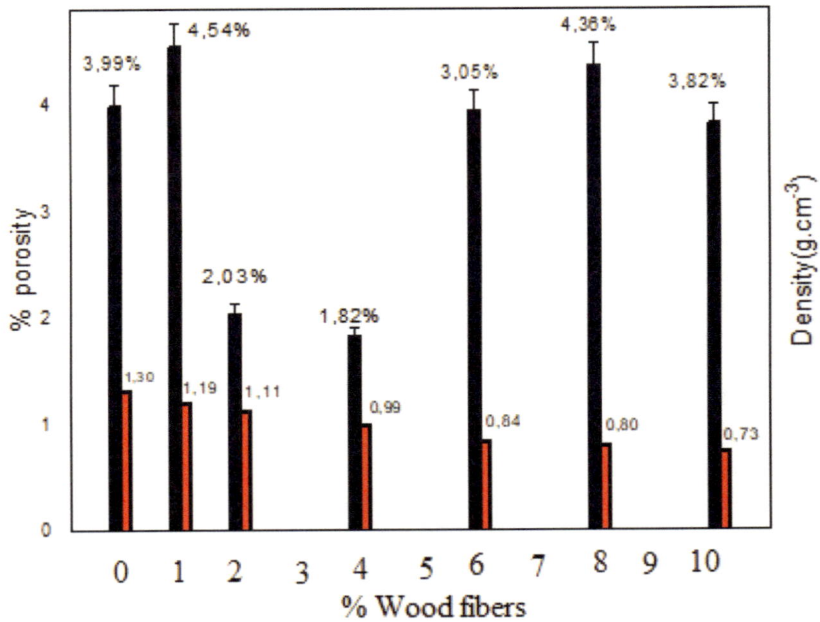

Figure 10.
Porosity and density as a function of treated wood fibers content for control cement and its composites.

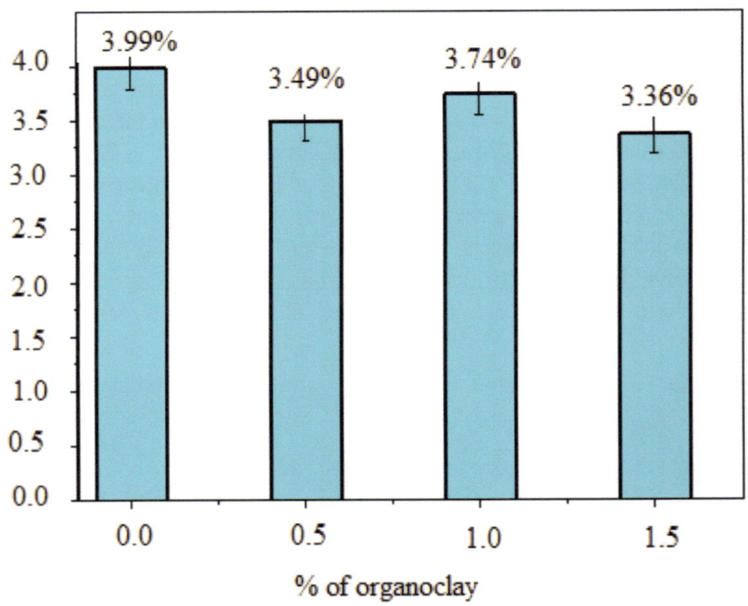

Figure 11.
Porosity as a function of organo clay content for control cement.

the porosity of composites by 18.75% when compared to the OPC. This implies that organo clay played a pore-filling role to reduce the porosity and to saturate pores. However, adding less than 1% wt of organo clay increased the porosity of all samples due to the agglomeration effect when less than 1% wt organo clay was added. This finding is comparable with the study where the porosity of the OPC is decreased due to the addition of 1% wt of organo clay to cement paste. We can say that composites containing 1% wt organo clay, are different from those of cement paste; the structure is less dense, and contain more pores. But composites with 1% wt of OC are different from those of OPC. The structure is more compact with few pores [25].

3.8 Mechanical studies

The incorporation of OC and WF with NaOH in to the cement composites improves the mechanical properties of the cement matrix. This addition leads to good improvement of the mechanical properties of samples [26].

3.8.1 Compressive resistance

- Effect of the Wood Fibers

The effect of WF treated by NaOH on the compressive strength of cement composite can also be seen evaluated. The compressive strength of WF reinforced cement composite is increased from 9,8 to 18.41 MPa about, 90% increases compared to treated WF reinforced cement composite. This enhancement improvement is explained as follows: To enhance the interfacial bond between fibers and the OPC, the matrix could be modified by reducing or consuming the calcium hydroxide. The low value of the compressive strength can be explained by the high sugar and hemicelluloses content of fast-growing wood. Conversely, the chemical treatment proved good mechanical properties.

- Effect of the Amount Organoclay

Effect of OC on the compressive strength of 0, 0.5, 1, and 1%wt. WFT-cement–composites of 28 days age are shown in histograms (**Figure 12**). It can be deduced that the compressive strength was increased by the addition of OC after (28 days) and the greater, the improvement. After 28 days, the compressive strength of 1%wt was 21.76 MPa higher than that of 0.5%wt of OC was 10.11 MPa. The reasons for the enhancement in mechanical properties of composites are as follows. Firstly, the physical effects of 1%wt of OC, including filling, can reduce the voids or the porosities in the cement matrix, in which the OC was uniformly dispersed in the matrix. Thus improving the microstructure of composites denser than the WFT-cement composite, Secondly, another mechanism is the pozzolanic reaction, in which OC reacts with calcium hydroxide (CH) in the cement matrix to produce calcium-silicate-hydrate (C-S-H). However, the addition of more than 1%wt of OC caused a marked decrease in compressive strength. For example, the compressive strength of 2% wt OC was 11.47 MPa. This is due to the poor dispersion and agglomeration of the OC in the cement matrix ata higher percentage of OC contents, which increase of porosity and reduce the bond strength between the fiber and the matrix adhesion. The compressive decreases in the cement-composite with a higher dosage of organo clay.

- Effect The Amount of SDBS

The effect of SDBS on compressive strength of composites with the non-treated wood fiber it shows that the addition of SDBS in cement matrix increased the compressive strength from 9.81 to 11.47 MPa, about 14.47% improvements can be

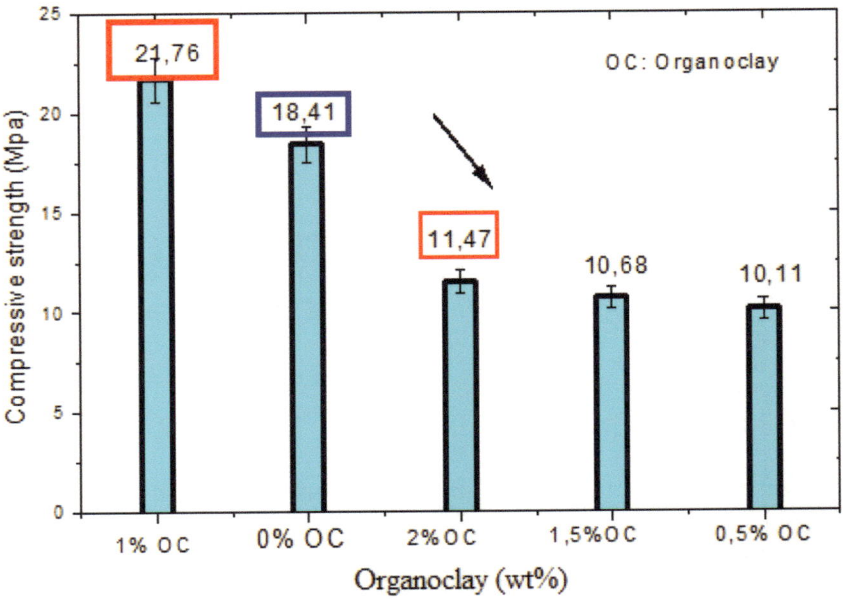

Figure 12.
Compressive strength as a function of organo clay content for control OPC with 6% wt.

explained by the adhesion and the collusion of fibers in the matrix in the presence of the SDBS.

The SDBS has an essential role in the surface packaging of the fiber; it is defined as a super plasticizer that covers the surface of the fibers and render them hydrophobic. In our case, anionic surfactant is able to agglomerate during the cement to fill the existing pores in the fibers.

3.9 Thermal conductivity

3.9.1 Effect of the number of fibers

The resulting curve shows that the thermal conductivity varies with the content of wood cement fibers. Decrease in the conductivity between 0 and 8% of the wood fiber. From 2.25 $Wm^{-1} K^{-1}$ for non-stabilizing blocks, it increases to 0.9 $Wm^{-1} K^{-1}$ when the wood fiber content is 2.5%. The thermal conductivity has increased for wood fiber contents ranging from 3–8%. For these wood fiber contents, the thermal conductivity ranges from 0.9 to 1.6 $W m^{-1} K^{-1}$.

The thermal conductivity of a material depends on several parameters such as the nature of the constituent elements of the material [27], the water content, the temperature and the porosity since the blocks are manufactured under the same conditions and the conductivity measurements are made in a stationary regime, the variation in conductivity can be related to the variation of porosity of the material, to the intrinsic composition of each sample and to the cohesion of the material. The decrease in thermal conductivity may be due to increased pore quantity or increased pore diameter due to poor distribution of wood fiber. In fact, it is considered that for this interval, the quantity of wood fibers is insufficient to favor the establishment of a homogeneous structure. Under the conditions of measurement of the thermal conductivity used, it is considered that the heat transfer takes place mainly by conduction. Thus, the pores represent a space in the transfer, an increase in their diameter or their quantity causes a slowing down of the heat transfer, hence the low measured thermal conductivity [28].

3.9.2 Effect of wood and Organoclay

At 30°C the study of 4 samples (1: Cement −2: C + WFNT, 3: C + WFT, 4: C + OC).

Generally, the thermal conductivity "λ" depends on the nature of the constituents, the temperature, the porosity and the water content. The decrease in thermal conductivity for these different samples could be related to the increase in the number of pores or the increase in pore diameter caused by poor and by poor distribution of cement. In this case, the heat transfer is by conduction only as well as the pores are shown gaps in the heat transfer since an increase in their diameter and their quantity causes the transfer of heat which confirms the weakness of the thermal conductivity.

An increase in thermal conductivity may be related to an increase in the temperature of the cohesion at different constituents of the sample.

Following the effect of cement hydration which allows the formation of portlandite and C-S-H which allow a strengthening of the bonds between the constituents that which promotes a decrease in porosity so we obtain a homogeneous structure; this structure is favorable for good heat transfer.

We note that the addition of 4% of the untreated fiber can decrease the thermal conductivity from 2.26 to 1.01. The **Figure 13** shows the addition of 1% OC causes a decrease in thermal conductivity from 1.39 to 1.08 W/m °C up to 22.30% this result proved that the addition of OC in the cementitious composite conducted to curb the heat exchange and to put the insulation capacity.

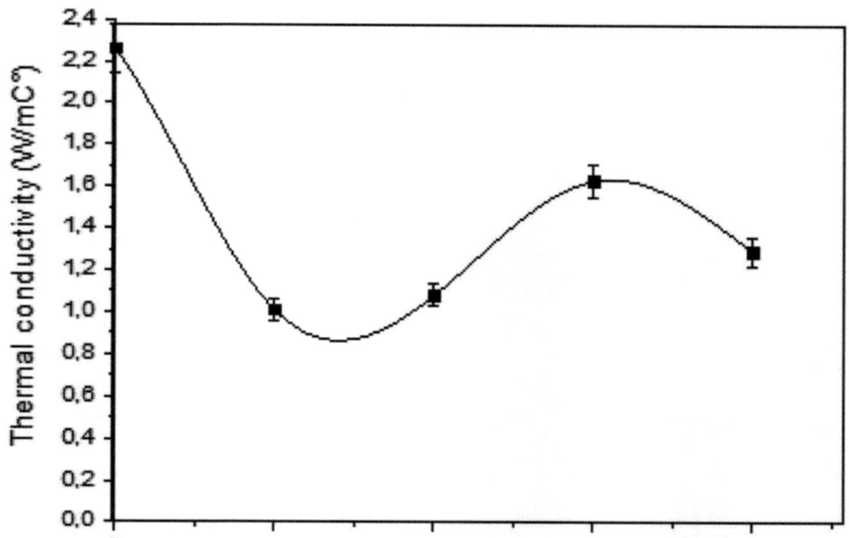

Figure 13.
Effect of amount of wood fiber on the thermalconductivity.

3.9.3 Effect of the porosity and the thermal conductivity

Figure 12 which examined the evolution of the porosity as a function of the percentage of addition, we observe that the addition of the wood fibers in the cement makes it possible to increase the porosity of this material (reinforcement by the natural and treated wood) by contrast, we notice that by adding Organo clay this porosity decreases and in this case we can conclude that the OC has just saturated the pores.

From this figure the thermal conductivity decreases as the percentage of fibers increase for reinforcement with 2 and 4%: this can be explained by the replacement of cement, an excellent natural insulator and a good thermal conductor. When the mass of the fibers reaches 8%, the thermal conductivity decreases in this case, the heat exchange by conductions weak in front of the convective exchange, as well as by the proportion of the pores which increases at the same time. Also in **Figure 14**, the variation of the thermal conductivity as the function of the mass of fibers and concerning the results, the thermal conductivity of cement materials reinforced with wood fiber and OC, showed a decrease with the increase of percentage in the mass of fiber in comparison with the reference sample. This reduction is essentially awed to the big porosity of the wood fiber and the big composite materials. To explain the decrease of different specimens, it can be concluded that 4%wt of wood fibers and OC contribute are to the thermal conductivity of the composites materials compared to the reference sample. Hence, incorporation of 6% wt of wood increase the thermal conductivity of the composite, but the thermal conductivity of hybrids composite is less than the reference.

Reasons for this type of behavior may be given as follows as: it has been found that the surface of the wood fiber exhibit the presence of the pore, which may reduce the adhesion of the fibers with the cement matrix, so we can say that the fibers-organo

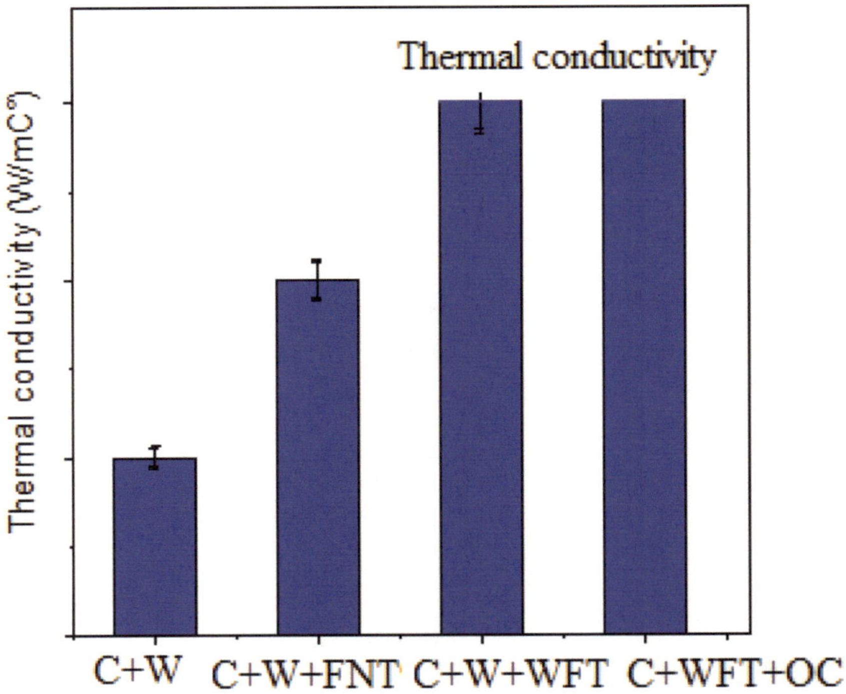

Figure 14.
Evolution of thermal conductivity as a function of additions (FT, FNT andOC).

clay–matrix adhesion plays an important role in the overall performance of the composite.

3.10 Interaction study between the fibrous suspension modified by an adjuvant and by the cement matrix

The interaction between the cement, wood fibers and clay modified by CTAB can be summarized in the following pictures **Figures 15** and **16**:

a. An adsorption of molecules having SO_3^{2-}, COO^- type functions or a polar function such as OH.

b. The interaction describes a phenomenon of an intergranular repulsion; in particularly is due to the case of superplasticizers, it is an adsorption of charged polymers.

c. The formation of micelles at the solid–solution interface.

d. This mechanism explain the chemisorption of polynaphthalene sulphonate on specific reaction sites such as these two aluminates.

e. This figure proposes the action of sugars or hydroxyl carboxylic acids by complexation in the interstitial solution. This complexation can then delay the precipitation of hydrates such as portlandite or C-S-H.

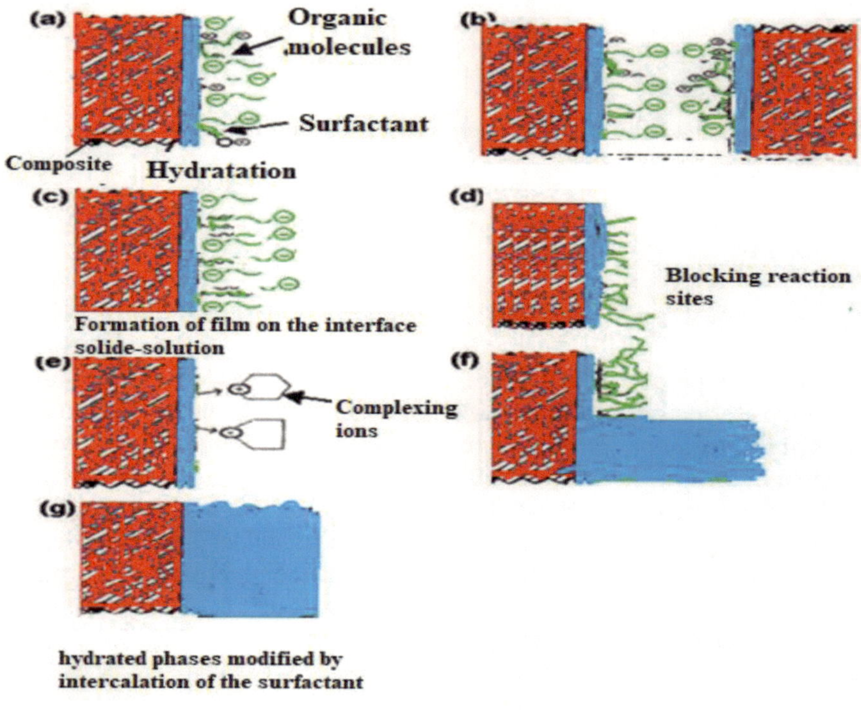

Figure 15.
Process describing the interaction between the composite and the surfactant.

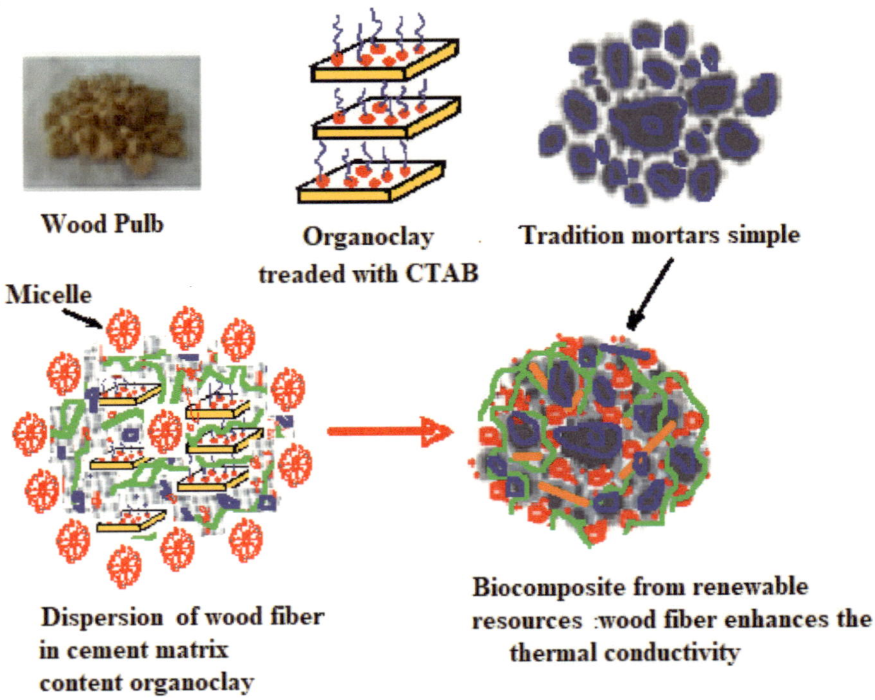

Figure 16.
Dispersion of WF and state of the environment of cement grains.

f. The mechanism (f) suggests that the CTAB surfactant have the role of potentially inhibiting the growth of hydrates by adsorbing on specific crystallographic growth sites.

g. The figure (g) describes the insertion of the polymer into the structure of the hydrate.

4. Conclusions

In this work, the wood fiber potential, as support for a cementitious matrix was examined.

The mechanical properties and the thermal conductivity of WFT/OC reinforced cement composite demonstrated the optimum content of organo clay is found to be 1%wt. For this situation, of composite, decrease in porosity (20%), thickness, improvement in compressive quality (20%) and a decline of thermal conductivity are improved by the expansion of 2%wt of wood.

Introducing OC and WFT in to the cement matrix would lead to a considerable increase in early age compressive strength and thermal conductivity. The compressive strength and the thermal conductivity of cement composites could be enhanced by the addition of treated wood fibers and organo clay. The optimum contents of OC and WFT are 1% wt and 6% wt respectively.

Moreover, introducing organo clay treated by the CTAB and wood fibers in to the cement matrix could lead to improve, accelerate hydration and reduce the thermal conductivity to ensure good insulation.

Acknowledgements

Authors are grateful to the Laboratory of Energy and Materials (LABEM), university of Sousse and Laboratoire Sciences des Matériaux et Environnement (LMSE), University of Sfax, Tunisia for the financial support of this work.

Conflict of interest

The authors declare no conflict of interest.

Notation

CTAB	Cetyltrimethylammonium bromide
CSH	Calcium-silica-hydration
NaOH	Sodiumhydroxide
OPC	Ordinary Portland cement NF P 15-301
OC	Organoclay
SDBS	Sodium dodecylbenzenesulfonate
W	Water
WFNT	Wood fiber nontreated
WFT	Wood fiberTreated

Author details

Fadhel Aloulou* and Habib Sammouda
Laboratory of Energy and Materials: LabEM-LR11ES34, University of Sousse, Tunisia

*Address all correspondence to: alouloufadhel@gmail.com

References

[1] Abdullah A, Jamaludin SB, Noor MM, Hussin K. Composite cement reinforced coconut fiber: Physical and mechanical properties and fracture behavior. Australian Journal of Basic and Applied Sciences. 2011;**5**(7):1228-1240

[2] Mehdizadeh P, Orooji Y, Amiri O, Salavati-Niasari M, Moayedi H. Green synthesis using cherry and orange juice and characterization of TbFeO3 ceramic nanostructures and their application as photocatalysts under UV light for removal of organic dyes in water. Journal of Cleaner Production. 2020;**252**:119765

[3] Assaedi H, Shaikh FUA, Low IM. Effect of nanoclay on durability and mechanical properties of flax fabric reinforced geopolymer composites. Journal of Asian Ceramic Societies. 2017;**5**(1):62-70

[4] Yang T, Wang Z, Tan S, Guo F. Mechanical Properties and Growth Mechanism of TiB2-TiC/Fe Composite Coating Fabricated in Situ by Laser Cladding. Applied Composite Materials. 2020;**27**(1):1-17

[5] Beji R, Hamdi W, Kesraoui A, Seffen M. Effects of salts on phosphorus adsorption in alkalize Tunisian soil. Euro-Mediterranean Journal for Environmental Integration. 2017;**2**(1):2

[6] Calabria-holley J, Papatzani S, Naden B, Mitchels J, Paine K. Applied clay science tailored montmorillonite nanoparticles and their behaviour in the alkaline cement environment. Applied Clay Science. 2017;**143**:67-75

[7] Chaallal O, Shahawy M, Hassan M. Performance of axially loaded short rectangular columns strengthened with carbon fiber-reinforced polymer wrapping. Journal of Composites for Construction. 2003;**7**(3):200-208

[8] Rodríguez-Sánchez J, Myszka B, Boccaccini AR, Dysthe DK. Setting behavior and bioactivity assessment of calcium carbonate cements. Journal of the American Ceramic Society. 2019;**102**(11):6980-6990

[9] Costa TCDC, Melo JDD, Paskocimas CA. Thermal and chemical treatments of montmorillonite clay. Ceramics International. 2013;**39**(5):5063-5067

[10] Durmaz S, Özgenç Ö, Boyaci IH, YIldIz ÜC, Erişir E. Examination of the chemical changes in spruce wood degraded by brown-rot fungi using FT-IR and FT-Raman spectroscopy. Vibrational Spectroscopy. 2016;**85**:202-207

[11] Fadhel A, Sabrine A. Preparation and evaluation of the influence of modified fiber flour wood on the properties of the fresh condition of cement - based mortars. International Journal of Industrial Chemistry. 2018;**9**(3):265-276

[12] El-Sokkary H, Galal K. Performance of eccentrically loaded reinforced-concrete masonry columns strengthened using FRP wraps. Journal of Composites for Construction. 2019;**23**(5):04019032

[13] Ghaffar SH. Straw fibre-based construction materials. Elsevier Ltd; 2016

[14] Goh S, Ku H, Ang SL. Prediction of crushing stress in composite materials. Journal of Composite Materials. 2008;**42**(5):467-484

[15] Hakamy A, Shaikh FUA, Low IM. Characteristics of hemp fabric reinforced nanoclay–cement nanocomposites.

Cement and Concrete Composites. 2014;**50**:27-35

[16] Hakamy A, Shaikh FUA, Low IM. Thermal and mechanical properties of NaOH treated hemp fabric and calcined nanoclay-reinforced cement nanocomposites. Materials & Design. 2015;**80**:70-81

[17] Hakamy A, Shaikh FUA, Low IM. Effect of calcined nanoclay on microstructural and mechanical properties of chemically treated hemp fabric-reinforced cement nanocomposites. Construction and Building Materials. 2015;**95**:882-891

[18] Humphreys MF. The use of polymer composites in construction. In: International Conference on Smart and Sustainable Built Environment. 2003;**1**(1):1-10

[19] Jalal M, Fathi M, Farzad M. Effects of fly ash and TiO2 nanoparticles on rheological, mechanical, microstructural and thermal properties of high strength self compacting concrete. Mechanics of Materials. 2013;**61**:11-27

[20] Norhasri MSM, Hamidah MS, Fadzil AM. Applications of using nano material in concrete: A review. Construction and Building Materials. 2017;**133**:91-97

[21] Parvin A, Wang W. Behavior of FRP Jacketed Concrete Columns under Eccentric Loading. 2001. pp. 146-152

[22] Alila S, Aloulou F, Beneventi D, Boufi S. Self-aggregation of cationic surfactants onto oxidized cellulose fibers and coadsorption of organic compounds. Langmuir. 2007;**23**(7):3723-3731

[23] Pacheco-Torgal F, Jalali S. Nanotechnology: Advantages and drawbacks in the field of construction and building materials. Construction and Building Materials. 2011;**25**(2):582-590

[24] Snoeck D, Smetryns PA, De Belie N. Improved multiple cracking and autogenous healing in cementitious materials by means of chemically-treated natural fibres. Biosystems Engineering. 2015;**139**(1998):87-99

[25] Styszko K, Nosek K, Motak M, Bester K. Comptes Rendus Chimie Preliminary selection of clay minerals for the removal of pharmaceuticals, bisphenol A and triclosan in acidic and neutral aqueous solutions´ raux d' argile pour l' e solutions aqueuses acides et neutres. Comptes Rendus Chimie. 2015;**18**(10):1134-1142

[26] Baley C, Davies P, Grohens Y, Dolto G. Application of interlaminar tests to marine composites. A literature review. Applied Composite Materials. 2004;11(77):98

[27] Sun J, Shi H, Qian B, Xu Z, Li W, Shen X. Effects of synthetic C-S-H/ PCE nanocomposites on early cement hydration. Construction and Building Materials. 2017;**140**:282-292

[28] Aloulou F, Alila S. Effect of modified fibre flour wood on the fresh condition properties of cement-based mortars. International Journal of Masonry Research and Innovation. 2019;**4**(4):355-377

Chapter 5

Oil Contaminated Sand: Sources, Properties, Remediation, and Engineering Applications

Rajab Abousnina and Rochstad Lim Allister

Abstract

Oil leakage during the exploration, production, and transportation of crude oil is a significant issue worldwide because crude oil spills severely impact the physical and chemical properties of the surrounding soil. A range of remediation methods for oil-contaminated soil is recommended, consisting of sand washing, bioremediation, electro-kinetic sand remediation, and thermal desorption; however, none are cost-effective. To find a suitable alternative remediation method, oil-contaminated sand utilisation in construction was considered. Several researchers found that oil contamination generally has an adverse effect on the mechanical properties of sand, but certain levels of contamination have beneficial effects on some of the important properties of the sand and its produced concrete. This chapter reviews the main sources of oil contamination and the existing remediation methods for this waste material. It analyses the different factors that affect the properties of oil-contaminated sand and concrete, including the type of crude oil and permeability of sand, like its properties, absorption, chemical composition, and spillage quantity. Furthermore, the intensive evaluation results of light crude oil effects on the geotechnical properties of fine sand, cement mortar and concrete were presented. Potential applications for oil-contaminated sand were also identified for the re-use of this material in engineering and construction.

Keywords: contaminated sand, produced water, remediation, mechanical properties, construction

1. Introduction

Sand contaminated with crude oil has become a major environmental concern worldwide. This problem poses threats to human health, the ecosystem, and the properties of the surrounding sand. Due to the prohibiting cost of the existing remediation methods, a more cost-effective way of utilising oil contaminated sand is warranted. Mixing oil contaminated sand with cement and using this mix as alternative construction material is considered an innovative approach to reduce its environmental impact.

There is growing public concern about the wide variety of toxic organic chemicals that are either deliberately or inadvertently being introduced into the environment.

Petroleum hydrocarbons are a common example of these chemicals because they frequently enter the environment in large volumes and in a variety of ways. Leakage from natural deposits is one way that crude oil affects the environment [1] and oil wastewater associated with the production of oil and gas is another source of oil contaminated sand [2–4]. Intentionally or accidentally, oil spill contamination impacts on the properties of the surrounding sand and changes its physical and chemical properties [2]. To minimise its effect on the environment, methods of remediation ranging from sand washing, bio-remediation, electro-kinetic sand remediation, and thermal desorption have been implemented, but are not considered to be cost effective [5]. One alternative method is using contaminated sand for engineering applications; indeed, some researchers have already investigated its use in that area and concluded that sand contaminated with oil can be used for road base materials or topping layers in parking areas [6–8]. However, the successful use of waste materials in concrete depends on the mechanical properties developed by the end product. While some studies investigated the effects of oil contamination on concrete, these studies only focused on heavy crude oil and engine oil [3, 9, 10] as well as hydrocarbons [10–12]. For instance, Almabrok et al. [13] investigated the effect that incorporated mineral oil has on the cement solidification process, and its consequent effect on the fresh and hardened properties of mortar. Almabrok et al. [14] further investigated oil solidification using a direct immobilisation method. Furthermore, the effect of kerosene impacted sand on the compressive strength of concrete in different conditions was investigated by Shahrabadi and Vafaei [15]. In this study, different percentages of kerosene (0, 0.5, 1, 2, 4, 6 and 8%, by weight of sand) was used to evaluate the effect of kerosene-soaked environment on compressive strength of concrete. They concluded that using contaminated sand adversely affects the compressive strength of concrete and hence, a reduction up to 27% in the concrete compressive strength occurred when samples with 2% of kerosene contamination were used. In a similar study by Attom et al. [16], the effect of 0.5, 1 and 1.5% of kerosene and diesel by dry weight of sand was investigated. They noted a reduction of up to 42% in the compressive strength as the level of contamination increased. Recently Shafiq et al. [17] investigated the effects of used engine oil (UEO) on the slump, compressive strength and oxygen permeability of normal and blended cement concrete. They concluded that engine oil caused variations in the compressive strength in the range of ±20% compared to the control mix. However, the UEO caused a reasonable reduction in total porosity and the coefficient of oxygen permeability of all concrete mixes compared to the control mix which can help this type of concrete achieve a high performance. Recently a study conducted by Abousnina et al. [18], the author investigated the effects of light crude oil contamination on the physical and mechanical properties of geopolymer cement mortar. The results showed that geopolymer mortar has the potential of utilising oil contaminated sand and reducing its environmental impact.

This shows the high potential of oil contaminated sand as sustainable material in building and construction. However, understanding the physical and mechanical properties of contaminated sand and its effect on produced mortar and concrete is very important in order to determine their end-use application [19]. Once achieved, this will potentially solve the issues of oil contamination in oil producing countries because the cost of this method will be cheaper compared to the existing remediation methods. Therefore, in this chapter the authors presented an extensive investigation to evaluate the effect that light crude oil has on the mechanical properties and microstructure of concrete. The succeeding sub-sections provide a better understating of the main sources, properties, existing remediation methods, and the beneficial use of oil contaminated sand in engineering and construction.

2. Petroleum contaminants

Over the last few decades there has been an increased public awareness of environmental issues, particularly when the contamination of sand, water, and air is involved. Worldwide, scientists and environmentalists are faced with the challenge of overcoming the detrimental effects of the contamination of sand, air, and water. The spillage of crude oil onto sand, leakages from pipelines, underground and surface fuel storage tanks, indiscriminate spills, and careless disposal and management of waste- and other by-products of society, constitute the major sources of petroleum contamination. **Figure 1** shows the final stage of produced water and the contaminated sand around the discharge disposal point, while **Figure 2** shows the oil spillage in one of the Libyan oil fields.

2.1 Sand contaminated with oil

One of the most critical environmental impacts of the oil industry is the spillage of crude oil, which severely contaminates the sand. Remediating contaminated sand takes longer and costs much more than it does for water contaminated with oil. Hence, it is very important to investigate the properties of sand contaminated with oil [21] because crude oil has a significant effect on the physical and chemical properties of surround sand [22]. However, the severity of the effect is based on several factors such as the sand type and quantity of the crude oil spillage. **Figure 3** shows who the crude oil migrates down to the groundwater causing partial saturation of the sand and the pathway of the hydrocarbons as shown in **Figure 3**.

Several studies have already investigated the effect of crude oil on the geotechnical properties of the sands. For instance, Cook et al. [24] experimentally investigated the compaction, compression, and strength properties of uniformly graded sands contaminated by crude oil. They reported that although oil contamination had no significant effect on the compaction characteristics, it decreases the friction angle and considerably increases the compressibility of the sand. Similar results were obtained by Puri [25] and Meegoda and Ratnaweera [26] for sandy and clayey

Figure 1.
Oil contaminated sand caused by produced water Ghani field HOO [20].

Figure 2.
Oil contaminated sand caused by oil spillage Ghani field HOO [20].

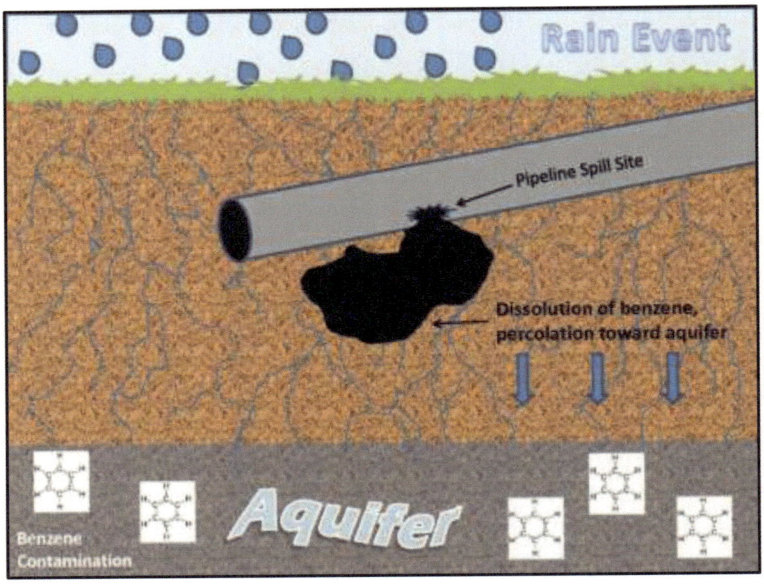

Figure 3.
Crude oil spillage and the leaching of light hydrocarbons to the aquifer [23].

sands, respectively. Studies on the geotechnical characteristics of fine-grained sands have just recently gained momentum. Khosravi et al. [2] studied the geotechnical properties of the oil-contaminated clayey and sandy sands and found a reduction in strength, permeability, maximum dry density, optimum water content, and Atterberg limits of these sands. Singh et al. [27] found an increase of 35–50% in the consolidation settlement of fine-grained sands upon contamination with petroleum hydrocarbons. From the previous studies in can be considered that the uniformly graded sands, clayey sands, fine-grained sands showed similar results however, when clayey and sandy sands was used the strength, permeability, maximum dry density, optimum water content, and Atterberg limits were decreased. Although none of the previous studies were investigated the absorption test of crude oil, it is expected that the type of sands may play a great role properties of oil contaminated sand.

Hydrocarbon contamination has a direct effect on the erodibility of sand and water infiltration and may also cause fire on the ground. Furthermore, the aggregation of fine particles, and the fusing of minerals, may lead to a reduction in the stability of the sand-organic matter aggregate. Whereas fire-induced or fire-enhanced sand water repellence has often been cited as the major cause of post-fire enhanced runoff and erosion [28], hydrocarbon contamination can also affect the physio-chemical characteristics of sand [29].

A previous study by Rahman et al. [30] has shown that is not only the ecosystem that can be affected by the spillage of crude oil, but the safety of the civil engineering structures as well. Cleaning up contaminated sand is a complicated job due to the long period of time needed, and the high cost and limitations of disposing the excavated sand. Furthermore, proper environmental regulations are not available due to the lack of proper management in many developing counties such as Libya, which then leads to disused oil and illegal dumping of other hydrocarbon components, which could have helped to tackle the environmental issue, as well as the economy, in the form of construction materials such as sand. Moreover, oil contamination can adversely affect plants as well as contaminate ground water resources for drinking or agricultural purposes [31].

2.2 Mechanical properties of oil contaminated sand

Sand is a naturally occurring material, which is considered to be an engineering material. Thus, its physical characteristics can be determined by experiments, which then may enable these properties to be used to predict their expected behaviour under working conditions, which in turn raises the possibility for its potential beneficial use to be determined [32]. The advantages of examining the mechanical properties of sand are that it permits a greater accuracy of measurements, so that any changes in conditions can be simulated to represent the conditions during and after construction, and the sand parameters can be derived within a reasonable time scale. Therefore, understanding mechanical properties of sand is beneficial to an engineer in terms of reducing the uncertainties in the analysis of foundations and earth work and the creation of structures, and in the use of sand as a construction material. In this application, the mechanical properties such as shear strength is investigated.

2.2.1 Shear strength

The shear strength of sand is one of the most important parameters in civil engineering applications. The safety of any mechanical engineering structure is based on the shear strength of the underlying sand [33]. All constructions, when in or on the land, impose loads on the sand that supports the foundations of that particular construction or building. The load imposed on the sand may cause shear failure of the underlying sand, which occurs when the shear stress imposed on the sand mass exceeds the maximum shear resistance (shear strength) that the sand can offer [34]. The shear strength of the sand is considered to be an important aspect in many foundations, such as in the bearing capacity of shallow foundations and piles, the lateral earth pressure on retaining walls, and the stability of slopes of dams [35]. Hence, understanding the shear strength can play a great role in terms of the entity classification of the sand [36], which in turn can assist engineers to derive the critical aspects of the overall sand mechanics in a specific environment.

The shear strength of common engineering materials, such as steel, from a continuum mechanics viewpoint, is governed by the molecular bonds that hold the material. The higher the shear strength of a material, the stronger the molecular structure [35]. Nevertheless, the shear strength of sand operates under a different set of principles. Sand is a particulate material, so shear failure occurs when the stresses between the particles are such that they slide or roll past each other. Due to the particulate nature of sand, unlike that of a continuum, the shear strength depends on the interaction of anti-particles (is it a particle of the opposite charge), rather than the internal strength of the sand particles themselves [36]. Sand derives its shear strength from two sources: cohesion between particles and frictional resistance between particles. Cohesion is the cementation between sand grains or the electrostatic attraction between sand particles [34].

2.3 Stabilisation of oil contaminated sand by mixing with cement

Several studies have investigated the mechanical properties of concrete utilising oil contaminated sand and have evaluated their potential use in construction. These studies are presented in this section, as well as the current usage of oil contaminated sand in construction.

2.3.1 Effects of oil contamination on the properties of mortar and concrete

Hamad and Rteil [37] revealed that oil acted like a chemical plasticiser, improved the fluidity, and doubled the slump of the concrete mix, while maintaining its compressive strength. A similar study was conducted by Hamad et al. [10] who added engine oil to a fresh concrete mix and found that its effect was similar to adding an air-entraining chemical admixture, which enhanced some of the durability properties of concrete. Additionally, the potential use of sand contaminated with petroleum in highway construction was investigated by Hassan et al. [38], and they concluded that it could be used for this purpose.

In a recent study by Ajagbe et al. [3] the effect of crude oil on compressive strength of concrete was investigated, and they concluded that 18–90% of its compressive strength was lost due to 2.5–25% contamination with crude oil. Ahad and Ramzi [9] indicated that there was a significant reduction in the compressive strength and about an 11% reduction in the splitting-tensile strength of concrete soaked in crude oil. **Table 1** summaries the effect of oil contamination on the mechanical properties of concrete. Nevertheless, there are still disagreements about the effect of crude oil and its produced content on the properties of produced concrete.

As it can be seen in **Table 1** that most of these researchers disagree on the effect of crude oil on the behaviour of concrete. The inconsistency of some of the factors such as type of crude oil, permeability of sand, sand properties, absorption, chemical composition, and spillage quantity [22, 39–41] were considered as the main reason beyond this lack of agreement. Thus, there is a need to further investigate the properties of oil contaminated sand and its effect on produced mortar and concrete.

2.3.2 Beneficial use of mortar and concrete containing oil contaminated sand

Range of remediation methods for sand contaminated with oil have been recommended but they are not cost effective [5]. The possibility of an end-use scenario of treated material or contaminated sand was addressed, based on the results of the compressive strength. For instance, less strength is required for landfill but a higher compressive strength is required to make bricks or for some other structural objectives. Based on the United States Environmental Protection Agency (USEPA)

Test	Petroleum hydrocarbon	Effect	References
Compressive strength	Engine oil	Maintained	[3]
	Oil, phenol, trichloroethylene (TCE) and hexachlorobenzene (HCB)	Decrease	[4]
	Hydrocarbon (phenol)	Maintained	[5]
	Low % of mineral oils	Maintained	[3]
	Crude oil	Decreased (18–90%)	[6]
	Soaked in crude oil	Decreased (11%)	[7]
Workability	Engine oil	Increased	[3]
Durability		Increased	

Table 1.
Effect of oil contamination on the properties of mortar/concrete properties.

guidelines, the recommended compressive strength, at 28 days, for landfill disposal material is 0.35 MPa, while it is 1.0 MPa in France and the Netherlands [42]. A higher compressive strength of 3.5 MPa in a sanitary landfill is required according to the Wastewater Technology Centre (WTC) in Canada [43]. Based on the British standard for precast concrete masonry units (BSI, 1981) a higher compressive strength is required for blocks and bricks, of 2.8 and 7 MPa, respectively. Additionally, a minimum of 7 days cube compressive strength, which should vary between 4.5 and 15 MPa, is required for subbase and base materials, as regulated under the department of transport in the UK. This shows that there is high potential in using oil-contaminated sand in construction. However, an understanding of the details of its physical and mechanical properties is warranted.

3. Results and discussion

3.1 Effect of light crude oil contamination on shear strength

The effect of oil contaminated sand with different percentages (0, 0.5, 1, 2, 4, 6, 8, 10, 15, and 20%) on shear strength was determined using the cohesion and frictional angle of fine sand contaminated with light crude oil, using the Mohr-Coulomb equation. **Figure 4** shows the sand shear strength as function of light crude oil contamination at an applied normal stress of 50 kPa. It can be deduced from the figure that light crude oil contamination affects the shear strength of fine sand.

The shear strength increased for fine sand with 1% oil contamination, which was due to the significant increase in the apparent cohesion of the fine sand contaminated with light crude oil [41]. In contrast, at a high level of crude oil contamination, the shear strength decreased significantly with a further addition of oil content beyond 1% of crude oil content in contaminated sand. Thus, it can be inferred that the presence of a high percentage of oil increased the viscosity, and as a consequence the sand particles started to be coated by the crude oil. By increasing the amount of oil content in particular sand, the chance of inter-particle slippage will also increase, and subsequently the shear strength of soil, will decrease [44]. This indicates that higher

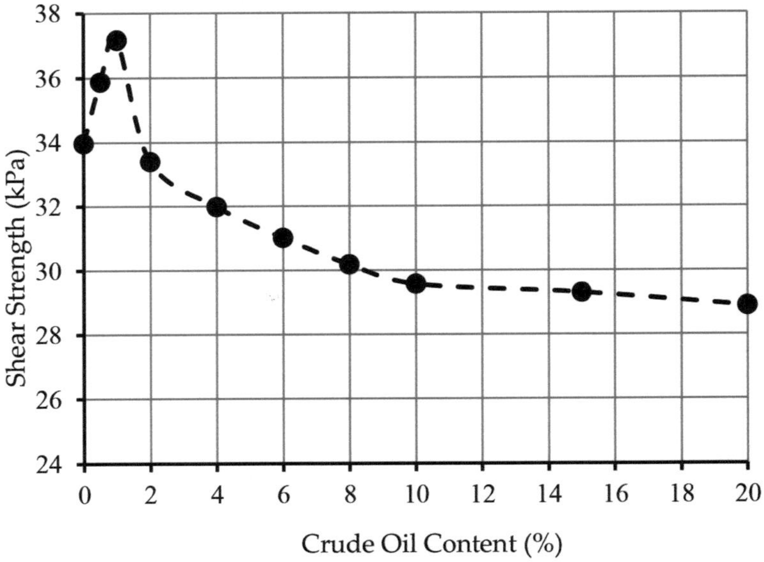

Figure 4.
Shear strength of contaminated fine sand at 50 kPa normal stress.

oil contamination resulted in lower shear strength values. Nevertheless, these results are contradicted by Khosravi et al. [2] finding, who concluded that the shear strength of kaolinite (clay) is not significantly influenced by gas oil. Furthermore, Puri [25], Shin and Das [45] both concluded that the shear strength of sand can be adversely affected by oil contamination. Moreover, Rahman et al. [44] have investigated the effect of hydrocarbons components on two types of soil, granitic soil and sedimentary soil. They concluded that the shear strength values of both soils significantly dropped from 0 to 4% of hydrocarbon content. Then, the shear strength values decreased slightly beyond 4% of hydrocarbon content. This implies that the sample will easily be slipped or sheared with higher oil content when the shear strain is applied. Moreover, Mashalah et al. [46] concluded in their study of the effect of crude oil contamination on three types of soils Lean clay, Silty sand and Poorly-graded sand (CL, SM and SP), that the shear strength decreased by increasing the crude oil content.

3.2 Effect of crude oil on the compressive strength of mortar and concrete

The effect of different percentages of light crude oil (0, 0.5, 1, 2, 4, 6, 8, 10, 15, and 20%) on the mechanical properties of mortar and concrete was investigated. A comparison between the compressive strength of cement mortar and concrete with different levels of light oil contamination is presented in **Figure 5**. In this figure, the percentage (%) increase or decrease in the compressive strength was calculated, based on the strength of the uncontaminated samples (0%). It is worth noting that the overall trend was similar with a higher value of the compressive strength, as obtained in mortar compared to concrete. This is despite the total volume of light crude oil being less for concrete than for cement mortar, due to the addition of coarse aggregates, even though the mix percentage of oil contamination by weight of sand is the same. At 1% of crude oil contamination, the cement mortar exhibited a 24.4% higher compressive strength than uncontaminated samples, while only an increase of 7.49% was shown by the concrete. At 2% of crude oil contamination, the mortar

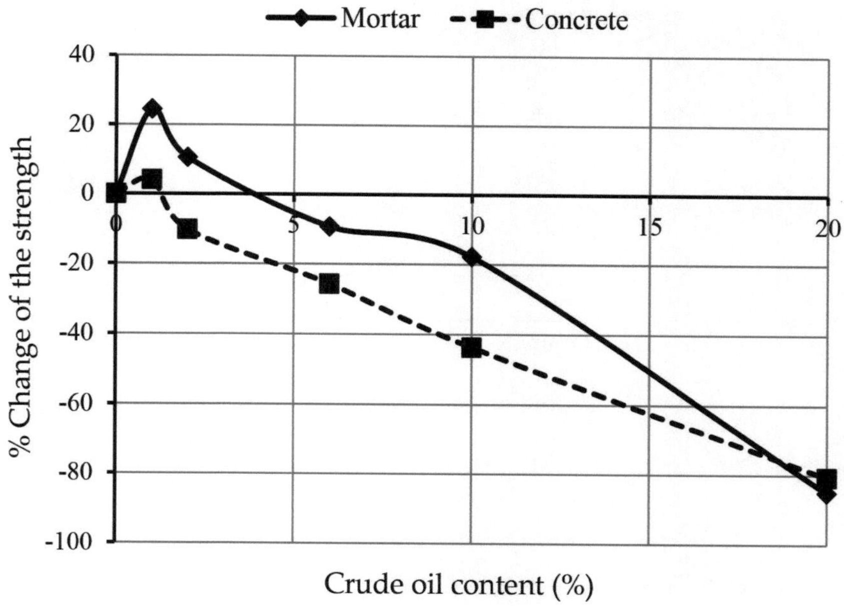

Figure 5.
Normalised compressive strength between mortar and concrete containing light crude oil contamination.

still has a 10.5% higher compressive strength than uncontaminated samples, but the compressive strength of concrete decreased by 10%. Up to 9 and 17% lower compressive strength was observed for mortar with 6 and 10% light crude oil contamination, respectively, whereas the decrease in concrete was as much as 25 and 43%, respectively. At 20% light crude oil contamination, the reduction of compressive strength was almost the same for both mortar and concrete, which was around 80%.

The higher reduction in compressive strength of the concrete with oil contamination, compared to cement mortar, was due to two main reasons: (1) lower cement binder to aggregates ratio (C: A); and (2) bigger aggregate size. Decreasing the amount of binder reduces the ability of the cement to fully cover the aggregates, hence yielding a void structure that forms a path for water permeation. It can be noted that the total volume of the fine sand used with cement mortar was around 75% (1:3), while the combination of fine and coarse aggregates for concrete accounted for almost 85% (1:3:3). This can be related to the internal structure of a pervious concrete, which is relatively less compact due to the insufficient amount of binders which produce the voids [47]. In addition, the presence of crude oil in high percentages hinders the bond formation between the cement to the large surface area of the coarse aggregates, resulting in more segregation in concrete than in mortar [48]. As a result, the compressive strength was lower for concrete compared to mortar. Furthermore, as the aggregate size increased, the water permeability of concrete increased. It is worth mentioning here that the connected porosity increased as the aggregate size increased. Previous studies conducted by Fu et al. [47], Chang et al. [49] showed that the compressive and splitting tensile strengths increased as the amount of binder increased, and they decreased with an increase in aggregate size.

The higher strength exhibited by the cement mortar than the concrete, under different percentages of crude oil content, may also be due to an increase in the contaminated surface area of the aggregates. In the concrete mix, the surface area of the

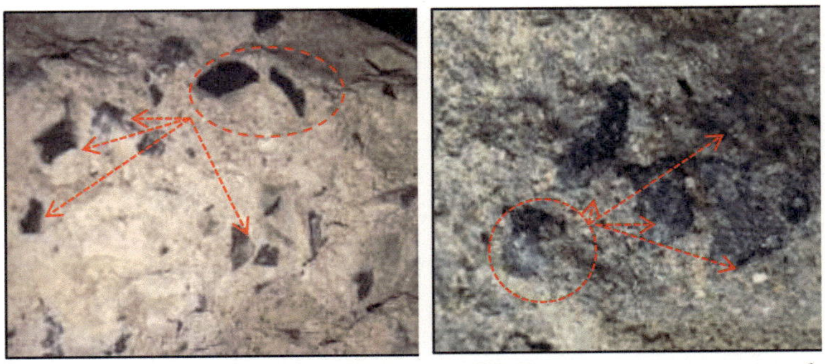

Figure 6.
Surface area of the coarse aggregate under different crude oil contaminations.

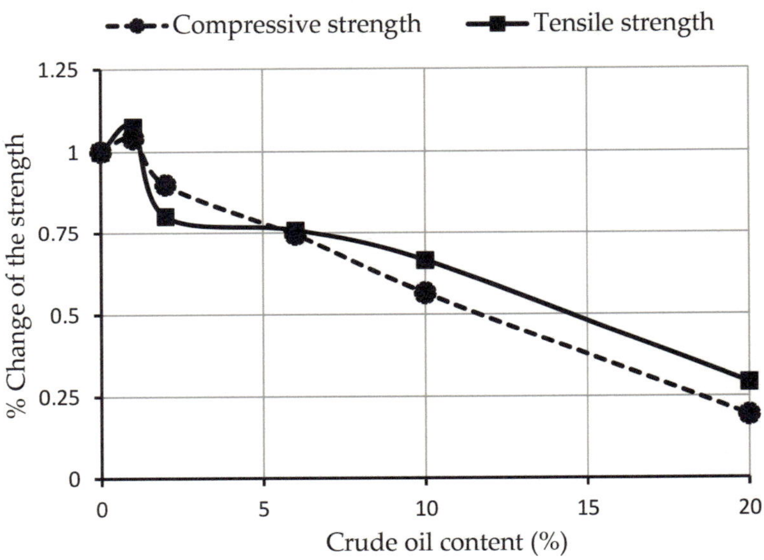

Figure 7.
Normalised relationships between the tensile compressive strength with different crude oil contaminations.

coarse aggregate is also coated by crude oil, as shown in **Figure 6**, especially at a high level of crude oil content of 10 and 20%. As a consequence, the total surface area of aggregate achieved damp or wet status (saturation status). Achieving the saturation status of aggregate by crude oil means extra free water will remain in the mix. This hindered the bond formation between cement and the coarse aggregates, resulting in more segregation for concrete than for mortar [48]. As a result, the compressive strength was lower for concrete compared to mortar.

3.3 Compressive and tensile strength relationship

Figure 7 shows the normalised effect on the compressive strength and splitting tensile strength of concrete containing fine sand with oil contamination. The behaviour of the splitting tensile strength shows a similar pattern to that of compressive strength, but with the percentage of reduction less than for compressive strength. Jasim [50] have attributed this behaviour to the inconsistency of the effect of crude oil contamination between the tensile and compressive properties. Kovler [51] has indicated that adding some waste materials, such as superabsorbent polymers (SAP), to concrete can improve the tensile strength to a higher extent than the compressive strength. This can be related to the fact that tensile strength is sensitive to cracking, so any improvement of cracking resistance is expected to be beneficial in terms of tensile properties. This indicates that the viscosity of oil had a significant effect on the splitting tensile strength of concrete, compared to its compressive strength.

4. Conclusion

The effect of crude oil on the mechanical properties of sand and its concrete were reviewed to evaluate their suitability for engineering applications and the following conclusions can be drawn:

- Sand contaminated by light crude oil is considered to be one of the most important environmental issues compared to medium and heavy crude oil because it contains a high percentage of light hydrocarbons that quickly spreads the crude oil into the surrounding soil.

- The highest value of cohesion was 10.76 kPa and it was observed at 1% of light crude oil contamination. This increase was due to the increased surface wetness of the fine sand particles. However, increasing the crude oil content from 2% up to 20% caused a significant reduction of cohesion value due to the sand particles being fully coated with crude oil, which reduced the interaction between sand particles.

- The compressive strength was enhanced by the crude oil up to the certain amount of crude oil level. At 1%, the oil contamination increased by 4.09% compared to the uncontaminated samples. However, the concrete containing fine sand with 2–6% of light crude oil contamination exhibited up to 25% lower compressive strength than uncontaminated samples and failed due to splitting failure. Increasing the crude oil from 10 to 20% resulted in significantly lower strength than the uncontaminated concrete.

- The optimum value of the splitting tensile was also observed at 1% of crude oil contamination, which was 7% higher compared to uncontaminated samples. On the other hand, increasing the crude oil content above 2% caused the fine sand to exceed the equilibrium condition, and the oil also contaminated the surface of the coarse aggregates.

- Oil contamination has a more detrimental effect on the mechanical properties of concrete than cement mortar. Up to 7 MPa lower compressive strength was obtained in concrete compared to mortar.

- Generally, it can be concluded that the properties of oil contaminated sand and its concrete were found to be suitable to use as some construction materials up to certain amount of level. Similarly, the results indicated that even the mortar and concrete using fine sand contaminated with higher crude oil contamination (15 and 20%) can be used for low load bearing engineering application such as landfill layering and production of bricks.

5. Proposals for future research

The following aspects need to be further studied in more detail to determine the end use application of cement mortar and concrete containing oil contaminated fine sand, and to pave the way for the development of structural components for civil engineering applications:

- The decrease in the mechanical properties of cement mortar and concrete containing oil contaminated fine sand could be enhanced by adding fibres into the mix. The appropriate forms of the fibres and the optimal dosage, which will result in properties equal or higher than the uncontaminated samples, need to be determined. Moreover, the maximum level of light crude oil contamination that will not severely impact on the bond between the cement and the fibres needs to be determined.

- The initial results of the effect of the light crude oil on the mechanical properties of geopolymer mortar showed a high potential to combine some of the major waste products, namely oil contaminated sand and fly ash, and to use them in building and construction. Therefore, further investigation on the effect of different percentages of oil contaminated sand on the mechanical and durability behaviour of the geopolymer concrete is recommended.

- The effect of crude oil on reinforced concrete is a new area of research that could be conducted to investigate the effect of crude oil on steel bars through the resistance of corrosion. Understanding the bond between reinforcement and concrete, and the behaviour of reinforced concrete structures containing oil contaminated sand warrants further exploration with regards to their potential beneficial use in civil engineering and construction applications.

Acknowledgements

Access to facilities at the Centre for Future Materials at USQ during specimen preparation and testing is acknowledged.

Conflict of interest

The authors declare that they have no conflict of interest.

DOI: http://dx.doi.org/10.5772/intechopen.103802

Author details

Rajab Abousnina[1*] and Rochstad Lim Allister[2]

1 School of Engineering, Faculty of Science and Engineering, Macquarie University, Sydney, NSW, Australia

2 TuffChem Environmental Services Pte. Ltd., Singapore, Singapore

*Address all correspondence to: rajab.abousnina@mq.edu.au

References

[1] National Academy of Science. Petroleum in the Marine Environment. Washington, DC: National Academy Press; 1975

[2] Khosravi E, Ghasemzadeh H, Sabour MR, Yazdani H. Geotechnical properties of gas oil-contaminated kaolinite. Engineering Geology. 2013;**166**:11-16

[3] Ajagbe WO, Omokehinde OS, Alade GA, Agbede OA. Effect of crude oil impacted sand on compressive strength of concrete. Construction and Building Materials. 2012;**26**(1):9-12

[4] Veil JA, Puder MG, Elcock D, Redweik Jr RJ. A white paper describing produced water from production of crude oil, natural gas, and coal bed methane. In: Prepared by Argonne National Laboratory for the US Department of Energy, National Energy Technology Laboratory. January 2004. Available from: http://www.ead.anl.gov/pub/dsp_detail.cfm

[5] Riser R. Remediation of Petroleum Contaminated Soils: Biological, Physical & Chemical Processes. United States: Lewis Publisher; 1998

[6] Virkutyte J, Sillanpää M, Latostenmaa P. Electrokinetic soil remediation: Critical overview. Science of the Total Environment. 2002;**289**(1-3):97-121

[7] Yeung AT, Gu Y-Y. A review on techniques to enhance electrochemical remediation of contaminated soils. Journal of Hazardous Materials. 2011;**195**:11-29

[8] Al-Rawas AA, Hago AW, Al-Sarmi H. Effect of lime, cement and Sarooj (artificial pozzolan) on the swelling potential of an expansive soil from Oman. Building and Environment. 2005;**40**(5):681-687

[9] Ramzi BA, Azad AM. Compressive and tensile strength of concrete loaded and soaked in crude oil. Engineering Journal of University of Qatar. 2000;**13**

[10] Hamad BS, Rteil AA, El-Fadel M. Effect of used engine oil on properties of fresh and hardened concrete. Construction and Building Materials. 2003;**17**(5):311-318

[11] Hebatpuria VM, Arafat HA, Rho HS, Bishop PL, Pinto NG, Buchanan RC. Immobilization of phenol in cement-based solidified/stabilized hazardous wastes using regenerated activated carbon: Leaching studies. Journal of Hazardous Materials. 1999;**70**(3): 117-138

[12] Cullinane MJ, Bricka RM, Francingues NR. An Assessment of Materials that Interfere with Stabilization/Solidification Processes, Report EPA/600/9-87/015. WA: Environmental Protection Agency; 1987

[13] Almabrok M, Mclaughlan R, Vessalas K. Effect of oil contaminated aggregates on cement hydration. 2013. pp. 81-89

[14] Almabrok MH, McLaughlan R, Vessalas K. Investigation of oil solidification using direct immobilization method. In: Presented at the Environmental Research Event 2011, North Stradbroke Island, QLD. 2011

[15] Shahrabadi H, Vafaei D. Effect of kerosene impacted sand on compressive strength of concrete in different

exposure conditions. Journal of Materials and Environmental Science. 2015;**6**(9):2665-2672

[16] Attom M, Hawileh R, Naser M. Investigation on concrete compressive strength mixed with sand contaminated by crude oil products. Construction and Building Materials. 2013;**47**:99-103

[17] Shafiq N, Choo CS, Isa MH. Effects of used engine oil on slump, compressive strength and oxygen permeability of normal and blended cement concrete. Construction and Building Materials. 2018;**187**:178-184

[18] Abousnina R, Manalo A, Lokuge W, Zhang Z. Effects of light crude oil contamination on the physical and mechanical properties of geopolymer cement mortar. Cement and Concrete Composites. 2018;**90**:136-149

[19] Abousnina RM, Lokuge W, Shiau J. Oil contaminated sand: An emerging and sustainable construction material. In: 2015 International Conference on Sustainable Design, Engineering and Construction; 10-13 May 2015; Chicago. 2015. pp. 1119-1126

[20] Rajab JS, Abousnina M, Manalo A, Lokuge W. Effect of light hydrocarbons contamination on shear strength of fine sand. In: Presented at the Fourth International Conference on Geotechnique, Construction Materials and Environment, Brisbane, Australia. 2014

[21] Jia Y, Wu Q, Shang H, Yang ZN, Shan H. The influence of oil contamination on the geotechnical properties of coastal sediments in the Yellow River Delta, China. Bulletin of Engineering Geology and the Environment. 2011;**70**(3):517-525

[22] Tuncan A, Pamukcu S. Predicted mechanism of crude oil and marine clay interactions. In: Proceedings of the Environmental Geotechnology. 1992. pp. 25-27

[23] NTSB. Hazardous Liquid Pipeline Rupture and Release. Washington, DC: NTSB; 2010

[24] Cook E, Puri V, Shin E. Geotechnical characteristics of crude oil-contaminated sands. In: The Second International Offshore and Polar Engineering Conference. 1992

[25] Puri VK. Geotechnical aspects of oil-contaminated sands. Soil and Sediment Contamination. 2000;**9**(4):359-374

[26] Meegoda NJ, Ratnaweera P. Compressibility of contaminated fine-grained soils. Geotechnical Testing Journal. 1994;**17**(1):101-112

[27] Singh S, Srivastava R, John S. Settlement characteristics of clayey soils contaminated with petroleum hydrocarbons. Soil & Sediment Contamination. 2008;**17**(3):290-300

[28] Shakesby R et al. Distinctiveness of wildfire effects on soil erosion in south-east Australian eucalypt forests assessed in a global context. Forest Ecology and Management. 2007;**238**(1):347-364

[29] Osuji L, Ezebuiro P. Hydrocarbon contamination of a typical mangrove floor in Niger Delta, Nigeria. International Journal of Environmental Science and Technology. 2006;**3**(3):313-320

[30] Rahman Z, Umar H, Ahmad N. Geotechnical characteristics of oil-contaminated granitic and metasedimentary soils. Asian Journal of Applied Science. 2010;**3**:237-249

[31] Rahman ZA, Hamzah U, Taha MR, Ithnain NS, Ahmad N. Influence of

oil contamination on geotechnical properties of basaltic residual soil. American Journal of Applied Sciences. 2010;**7**(7):954

[32] Head KH. Manual of Soil Laboratory Testing. Volume 1: Soil Classification and Compaction Tests. 3rd ed. Vol. 1. Caithness: Whittles Publishing; 2006

[33] Budhu M. Foundations and earth retaining structures. John Wiley & Sons Incorporated; 2008

[34] Smith I. Smith's Elements of Soil Mechanics. Wiley; 2013

[35] Das BM. Advanced Soil Mechanics. Psychology Press; 2008

[36] Coduto DP, Kitch WA, Yeung M-CR. Foundation design: Principles and practices. Vol. 2. USA: Prentice Hall; 2001

[37] Hamad BS, Rteil AA. Effect of used engine oil on structural behavior of reinforced concrete elements. Construction and Building Materials. 2003;**17**(3):203-211

[38] Hassan HF, Taha R, Al Rawas A, Al Shandoudi B, Al Gheithi K, Al Barami AM. Potential uses of petroleum-contaminated soil in highway construction. Construction and Building Materials. 2005;**19**(8):646-652

[39] Nudelman N, Ríos S, Katusich O. Organic co-solvent effect on the estimation of the equilibrium aqueous concentrations of oil residuals in Patagonian soil. Environmental Technology. 2002;**23**(9):961-970

[40] Fine P, Graber E, Yaron B. Soil interactions with petroleum hydrocarbons: Abiotic processes. Soil Technology. 1997;**10**(2):133-153

[41] Abousnina RM, Manalo A, Shiau J, Lokuge W. Effects of light crude oil contamination on the physical and mechanical properties of fine sand. Soil and Sediment Contamination: An International Journal. 2015;**24**(8): 833-845

[42] Spence RD, Shi C. Stabilization and Solidification of Hazardous, Radioactive and Mixed Wastes. Boca Raton, Florida, USA: CFC Press; 2005. ISBN: 1-56670-444-8

[43] Stegemann J, Cote P. A proposed protocol for evaluation of solidified wastes. Science of the Total Environment. 1996;**178**(1):103-110

[44] Rahman Z, Umar H, Ahmed N. Geotechnical characteristics of oil contaminated granitic and metasedimentary soils. Asian Journal, Applied Science. 2010;**3**:273-249

[45] Shin EC, Das BM. Some physical properties of unsaturated oil-contaminated sand. In: Advances in unsaturated geotechnics, Geo-Denver. 2000. pp. 142-152. DOI: 10.1061/40510(287)9

[46] Mashalah K, Amir H, Majid T. The effects of crude oil contamination on geotechnical properties of Bushehr coastal soils in Iran. In: The Geological Society of London. IAEG2006 Paper Number. Vol. 214. 2006

[47] Fu TF, Yeih W, Chang JJ, Huang R. The influence of aggregate size and binder material on the properties of pervious concrete. Journal of Advances in Materials Science and Engineering. 2014;**2014**:17. Article ID-963971. DOI: 10.1155/2014/963971

[48] Osuji S, Nwankwo E. Effect of crude oil contamination on the compressive strength of concrete. Nigerian Journal of Technology (NIJOTECH). 2015;**34**(2):259-265

[49] Chang J, Yeih W, Chung T, Huang R. Properties of pervious concrete made with electric arc furnace slag and alkali-activated slag cement. Construction and Building Materials. 2016;**109**:34-40

[50] Jasim AT, Jawad FA. Effect of oil on strength of normal and high performance concrete. Al-Qadisiyah Journal for Engineering Sciences. 2010;**3**(1):24-32

[51] Kovler K. Effect of superabsorbent polymers on the mechanical properties of concrete. In: Mechtcherine V, Reinhardt H-W, editors. Application of Super Absorbent Polymers (SAP) in Concrete Construction: State-of-the-Art Report Prepared by Technical Committee 225-SAP. Dordrecht, Netherlands: Springer; 2012. pp. 99-114

Chapter 6

The Data Representations of a Building Project: BIM Model, and IFC or IFCXML Data Standard

Murat Aydın

Abstract

Building regulations in the construction industry are legal documents written in human language. These are interpreted and implemented by people and generally controlled by local governments. Traditional building regulation control and supervision methods emerge as a time-consuming and error-prone process for architects, engineers, and public authorities. Therefore, BIM's effective building regulation control is considered a promising field of study in the construction industry. Automated Code Compliance Checking (ACCC) method is a rule-based method that provides simultaneous control of the computer's building regulations. ACCC takes into account the characteristics of the building elements and related building regulations. BIM is recognized as the most effective platform for information exchange of building projects in the construction industry. It supports the development of various software. It facilitates automated or semi-automated ACCC of the building projects for compliance with building regulations and standards for the participants involved in the building production process. The data of the building project are represented in two ways in the ACCC. These are BIM Model, and IFC or IFCXML Data Standard. In this study, the BIM, IFC, and IFCXML representations of the building project data were explained over the sample housing project in the ACCC process.

Keywords: BIM, IFC, IFCXML, housing project, automated code compliance checking

1. Introduction

Significant advances have been made in building design and construction with technology development in the construction industry. Before constructing the building project, which was drawn on paper, scale physical models are created following the building design through the developing technology. Better quality buildings are started to be built in less time and cost than planned with these models. The use of BIM (Building Information Modeling) for the physical models has been an essential step for building designs in the construction industry before starting the construction of building projects. There has also been an increase in the use of information technologies with BIM, such as saving data, exchanging data, and verifying the conformity of data according to building regulations among the building project life cycle. Optimum solutions have begun to be obtained with informatics to facilitate the creation, transformation, and use of information to increase the automation of paper-based manually controlled processes in the construction industry [1–4].

Building regulations in the construction industry are legal documents written in human language. These are interpreted and implemented by people and are generally controlled by local governments. Traditional building regulation control and supervision methods emerge as a time-consuming and error-prone process for architects, engineers, and public authorities [5–7]. Therefore, BIM's effective building regulation control is considered a promising field of study in the construction industry. Automated Code Compliance Checking (ACCC) method is a rule-based method that provides simultaneous control of the computer's building regulations. ACCC takes into account the characteristics of the building elements and related building regulations. In this method, each building element is checked for compliance with the rules and conditions of the relevant building regulation. Then, it creates a result report related to the building elements [8–10]. Shortly, ACCC is the answer to how the computer can interpret the building regulations, the computer can create the building regulation rules, and the automated code compliance checking of the building project can be controlled by the computer according to the building regulations [11].

2. Method

BIM is recognized as the most effective platform for information exchange on building projects in the construction industry. It supports the development of various software. It facilitates automated or semi-automated ACCC of the building projects for compliance with building regulations and standards for the participants (designer, architect, engineer, contractor, owner, etc.) involved in the building production process. The data of the building project are represented in two ways in the ACCC [12]. These are:

- BIM Model, and
- IFC or IFCXML Data Standard.

2.1 BIM model

Building Information Modeling (BIM) is a simulation prototyping technology. The definition of the BIM and BIM model according to the US National BIM Standard is "BIM is the digital representation of the physical and functional characteristics of a building project. The BIM model is a reliable source of information where information is shared from conceptual design to demolition of the building project throughout the building project life cycle [13]." BIM, one of the most critical developments in the construction industry, emerges as a technology that enables the management of the building project data in a digital environment by incorporating different tools and processes into the building design. It is based on the building elements that form the building project. It models the relationships of the building elements with each other. Supporting interdisciplinary integration in the construction industry, BIM effectively changes its role in building design and construction by creating a database of the building elements used throughout the building project life cycle [14, 15].

2.2 IFC data standard

Industry Foundation Classes (IFC) data standard, a new industry foundation class for interoperability, was created by the Industry Alliance for Interoperability (IAI) in 1997 [16]. IFC is an object-oriented data standard developed in the EXPRESS language, and it is independent of any software [17]. IFC is supported by BIM-based software. Therefore, it is accepted that BIM and IFC data standards will significantly advance and facilitate cooperation in the building design process. IFC provides users with comprehensive information and features about the building project. It also represents international standardized object definitions [18]. Objects are also named as the building elements in the building project. The most important feature of the IFC data standard is its rich data structure that allows a building element to be defined by more than one feature. The IFC standard data is divided into different building elements such as the wall, column, beam, floor, window, door, railing, elevator, stair, etc. These building elements have a three-dimensional geometry and parameters, which can be divided into size, material, property, price, quantity, etc.

2.3 IFCXML data standard

IFCXML is a language defined in XML, equivalent to the EXPRESS-based specification of IFC data. The implementation of the IFC data standard using XML is named IFCXML. IFCXML is the ".ifcxml" extension of the current IFC data format. It is an implementation of the ISO 10303-28 standard [18]. This standard provides an automatic conversion feature from IFC's EXPRESS language to an XML language. By representing IFC data in XML language, users can perform many operations such as extracting, transferring, using, and merging IFC data between various applications. The IFCXML representation provides for ease of understanding classes in IFC along with sub-classes. This feature simplifies the use of IFC by reducing the complexity of IFC data. Due to the XML feature, the IFCXML file size is larger than the IFC file size (approximately five times).

3. Sample housing project

In this study, a sample housing project was prepared for the BIM, IFC, and IFCXML representation of the building project data in the ACCC process. The sample housing project consisted of 9 floors, 8 flats, and 2 elevators. Flat plans on each floor were modeled differently from each other. It was completed with 179 rooms, 91 doors, 66 windows, 52 railings, 2 mechanical ventilation, 2 elevators, 8 stairs, 92 walls, 135 columns, 183 beams, and 95 slabs. As shown in **Figure 1**, the procedure for creating the sample housing project with ArchiCAD software are as follows:

- Creating a new building project,
- Entering the building project information,

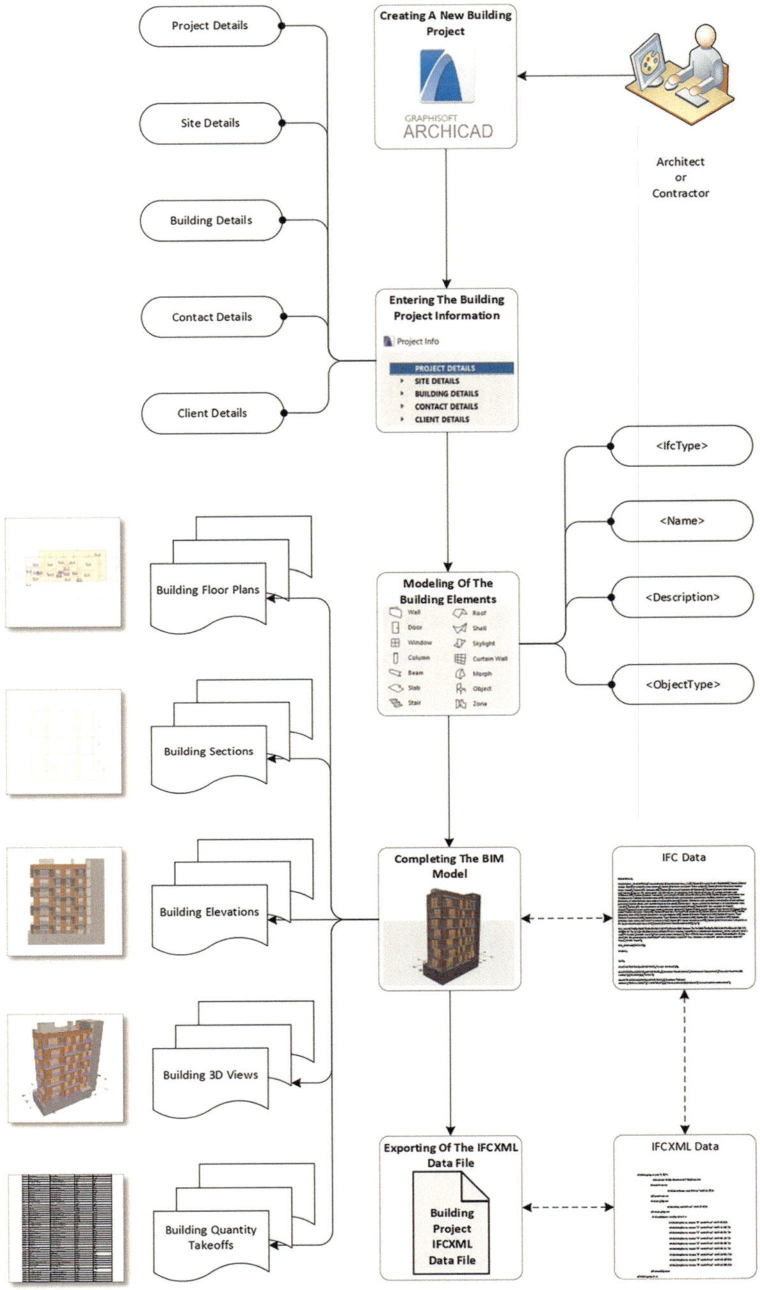

Figure 1.
The procedures in creating the sample housing project with ArchiCAD software [19].

- Modeling of the building elements,
- Completing the BIM model, and
- Exporting of the IFCXML data file.

3.1 Creating a new building project

As shown in **Figure 2**, a new building project named "Sample Housing Project" was created by clicking the New sub-menu under the File menu in ArchiCAD. The default template set was selected for the template option of the new building project in ArchiCAD. First, the working unit settings were set in the sample housing project. "1/ 50" for the scale unit, "cm" for the length measurement unit, "m2" for the area measurement unit, and "m3" for the volume measurement unit was selected in the working unit settings in ArchiCAD. After the working unit settings were set, the sample housing project's floor information was named respectively -1. Floor, 0. Ground Floor, 1. Floor, 2. Floor, 3. Floor, 4. Floor, 5. Floor, 6. Floor and 7. Floor.

3.2 Entering the building project information

As shown in **Figure 3**, the Project Info option in the Info sub-menu under the File menu in ArchiCAD was selected to add the sample housing project's project information. The project information was entered under five headings on the Project Info page automatically provided by ArchiCAD software. These headings are detailed below:

- Project details,
- Site details,
- Building details,
- Contact details, and
- Client details.

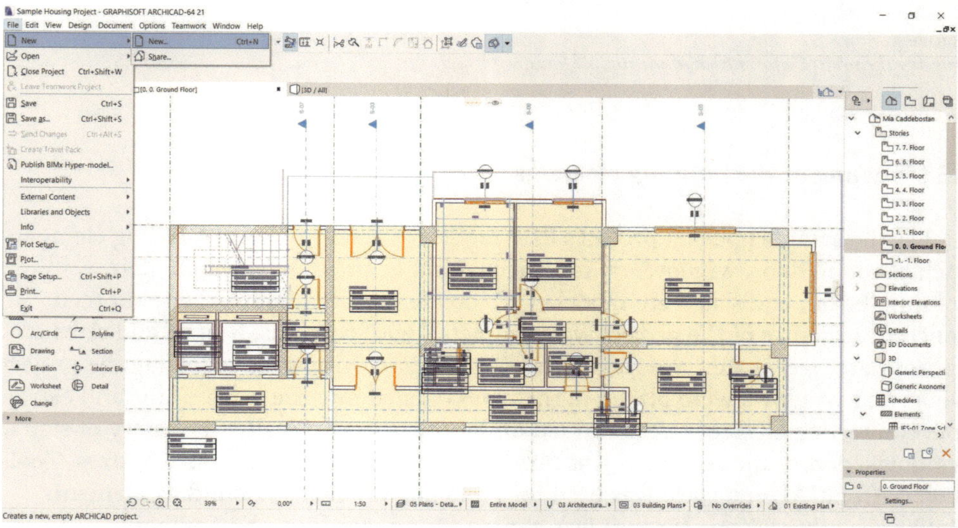

Figure 2.
Creating a new housing project in ArchiCAD.

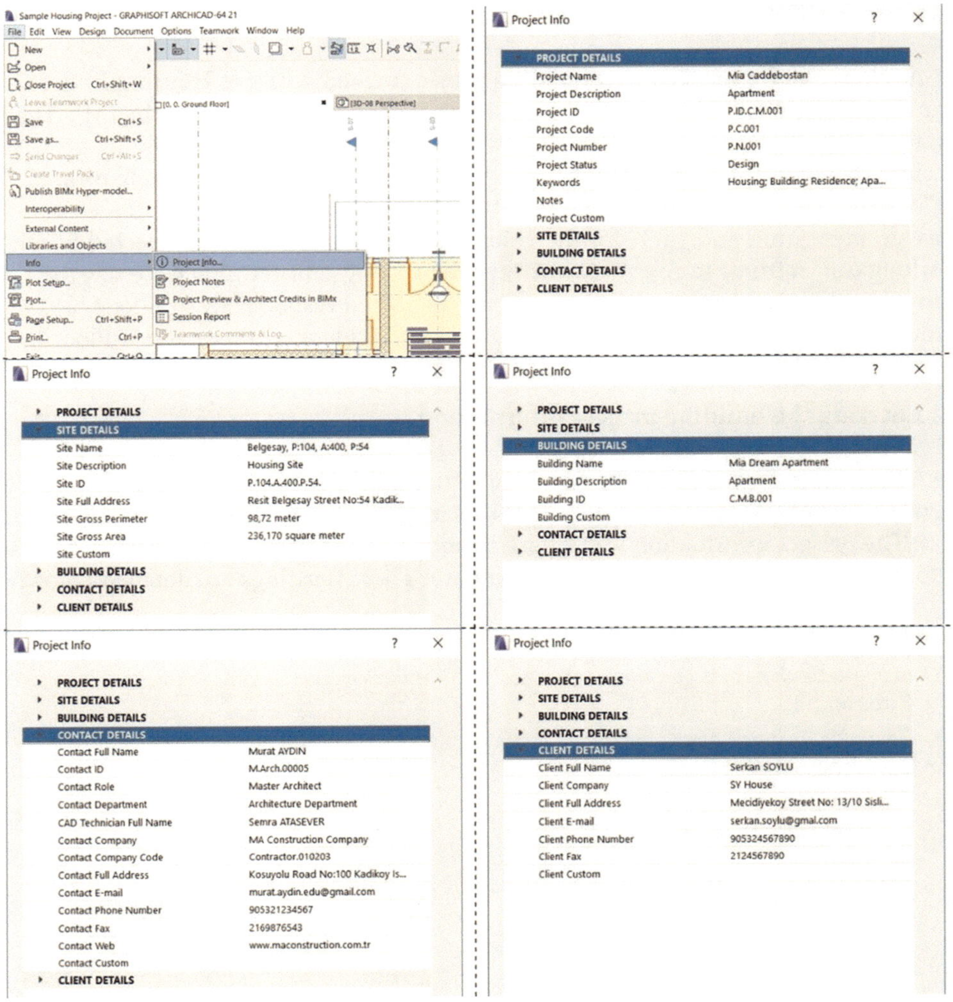

Figure 3.
Entering the project information of the sample housing project.

3.3 Modeling of the building elements

For the sample housing project's BIM model, different types of building elements such as room, door, window, railing, mechanical ventilation, elevator, stair, wall, column, beam, and slab were modeled. To create the BIM model, the reinforced concrete structure of the project was completed with columns, beams, and slabs of appropriate sizes. Later, the rooms were determined with walls following the project. Doors and windows were placed in the holes opened into the walls. Stairs and elevators connecting all floors were formed under the project. Finally, the safety of stairs, windows, doors, and balconies was provided with railings in appropriate sizes. Toolbox in ArchiCAD interface was used for modeling the selected building elements. Unlike the others, Zone (Space) was chosen for a room from the Toolbox.

As shown in **Figure 4**, the <IfcType> and <Name> attributes of the building elements were assigned automatically by ArchiCAD. The <IfcType> attribute is not allowed to be changed by the user. However, the <Name> attribute can be changed by the user if required. ArchiCAD gave the <Name> attribute of the building element by automatically associating it with the floor name and sequence number. For example, WD – 003 Living Room Window shown in **Figure 4** is the third window ('WD' – 0'03') on the ground floor (WD – '0'03) of the sample housing project. That is why it

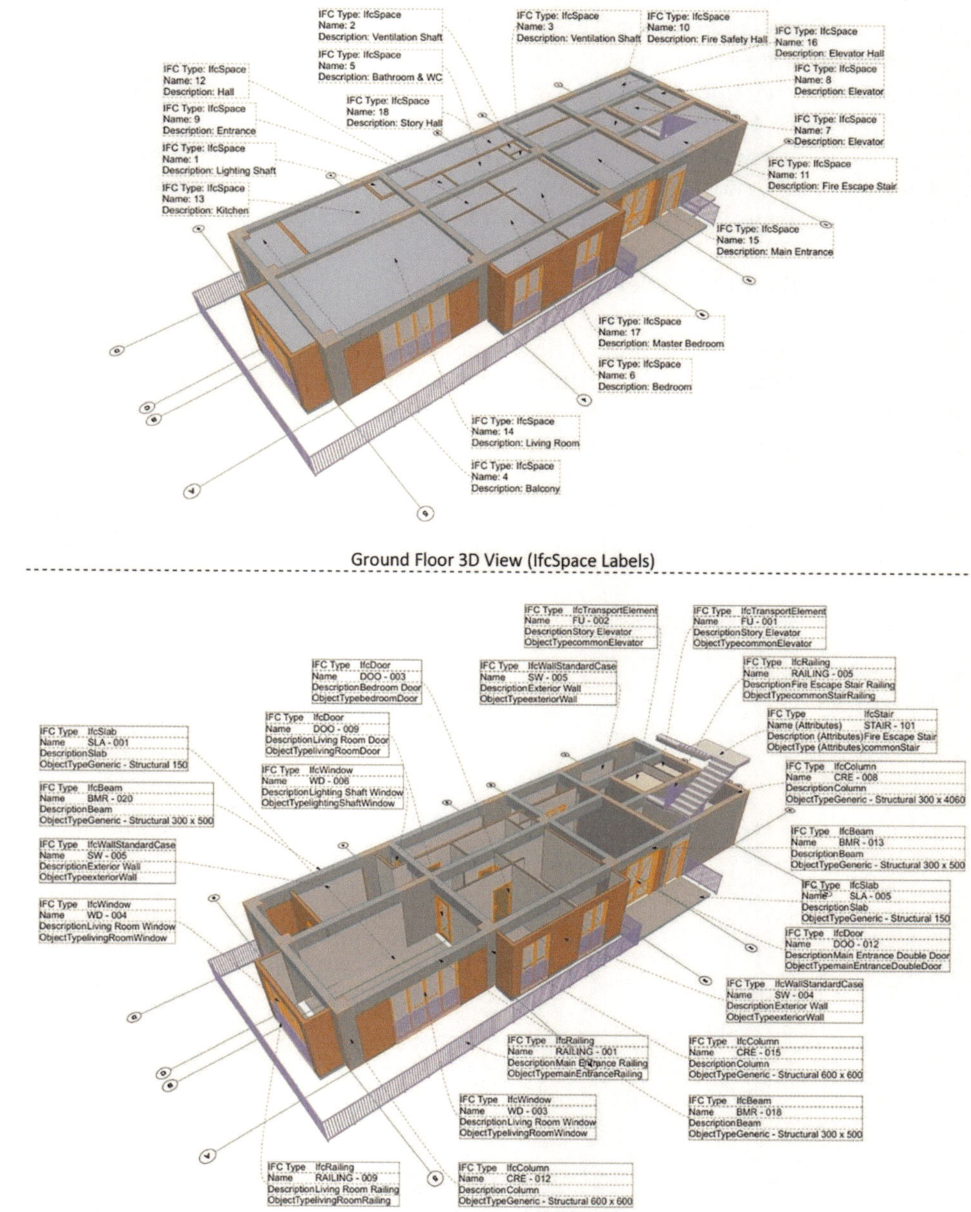

Ground Floor 3D View (IfcDoor, IfcWindow, IfcRailing, IfcBuildingElementProxy, IfcTransportElement, IfcStair, IfcWallStandartCase, IfcColumn, IfcBeam and IfcSlab Labels)

Figure 4.
The representation of the IFC data file of the sample housing project (the building elements on the ground floor with IFC labels).

is named WD – 003. The <Description> and <ObjectType> attributes of the building elements were entered by the user. The <Description> attribute is the appropriate selection of the relevant building element for the room to which it is associated. On the other hand, the <ObjectType> attribute is the programming language equivalent of selecting the relevant building element, which is appropriate for the room to which it is associated.

An example of the building elements whose four attributes were defined with the IFC labels was shown in **Figure 4**. Three-dimensional views of the building elements on the ground floor were automatically prepared with ArchiCAD. The building elements were shown with the <IfcType>, <Name>, <Description>, and <ObjectType> attributes in **Figure 4**. The room (space) labels of the building elements on the BIM model's ground floor were shown in the top view. The window, railing, mechanical ventilation, elevator, stair, wall, column, beam, and slab labels of the building elements on the BIM model's ground floor were shown in the bottom view.

3.4 Completing the BIM model

In this study, the working unit settings were set first. Then, the floor information of the sample housing project was named. The project information was entered under five headings; different types of building elements were modeled. Thus, the BIM model of the sample housing project was completed with all these procedures. The

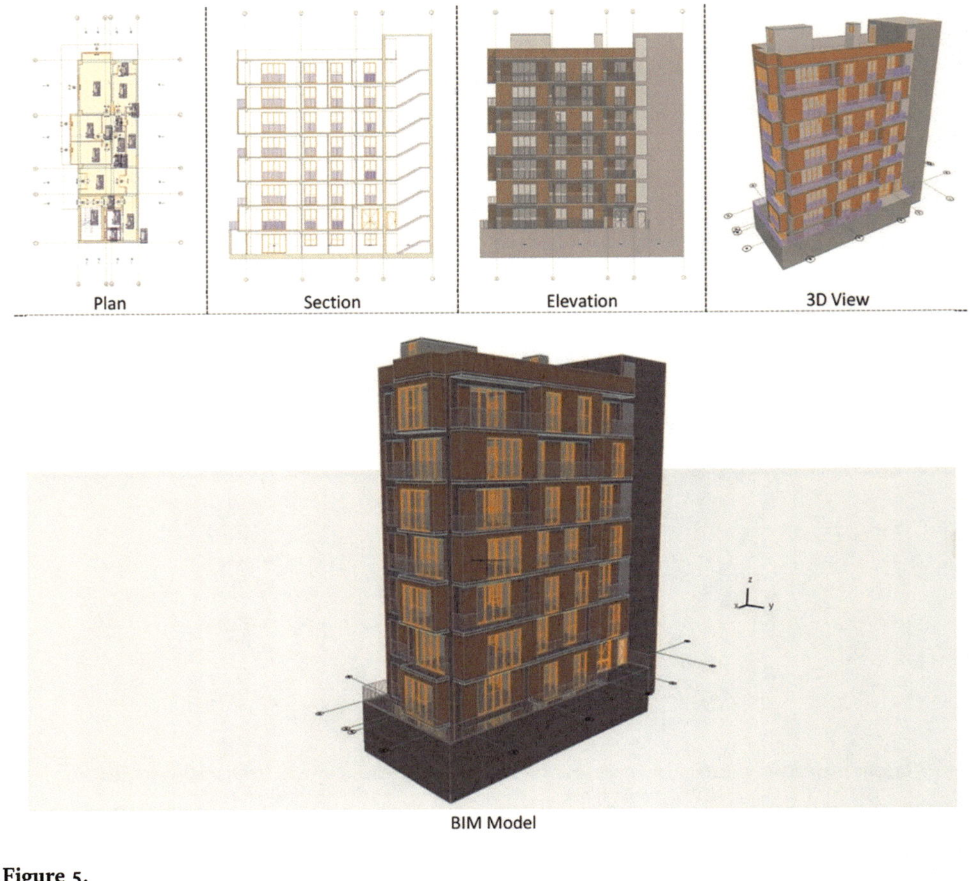

Figure 5.
The representation of the BIM model of the sample housing project.

BIM model consisted of 9 floors, 8 flats, and 2 elevators. Except for the 7. Floor, which was shaped as a terrace, each floor was designed as one flat. Apart from the vertical circulation, mechanical ventilation, and installation spaces, each floor's flat plans were formed differently from the others. Finally, the representation of the sample housing project's BIM model was shown as a three-dimensional view in **Figure 5**.

As seen in **Figure 5**, the sample housing project's project documents were automatically prepared with ArchiCAD software. The project documents of the project were printed with the default publisher settings in ArchiCAD. These are:

- Building floor plans,
- Building sections,
- Building elevations,
- Building 3D views, and
- Building quantity takeoffs.

3.5 Exporting the IFCXML data file

The IFCXML data file was exported with the completion of the BIM model of the sample housing project. The Save As Type option was selected as IFCXML, and the Export option was selected as Entire Project by clicking the Save As sub-menu under the File menu in ArchiCAD. After selection, the IFCXML data file of the sample housing project was exported. The first page of the IFCXML data file consisting of approximately 5.700.000 strings is shown in **Figure 6**.

In IFCXML, the data starts with the "Ifc" prefix, just like in the IFC data representation. According to the CamelCase naming convention, the first letters are

Figure 6.
The representation of the IFCXML data file of the sample housing project.

capitalized, and English words are without underlines. The word must be shown in a string (< ... .>). Each class has an ID number in a string. ID number is given mainly for each class, and it is specified in the string with ref = " ... .". ID numbers are required to provide, clarify and verify relationships between classes. The logical relationship between IfcProject and IfcBuildingStorey of the sample housing project is shown in the IFCXML language in **Figure 7**.

A superclass matches one or more subclasses with <IfcRelAggregates> in IFCXML. The superclass is specified with <RelatingObject> and its number is always one. Therefore, it is not represented by the ex: pos = " ... " number. Subclasses or classes are

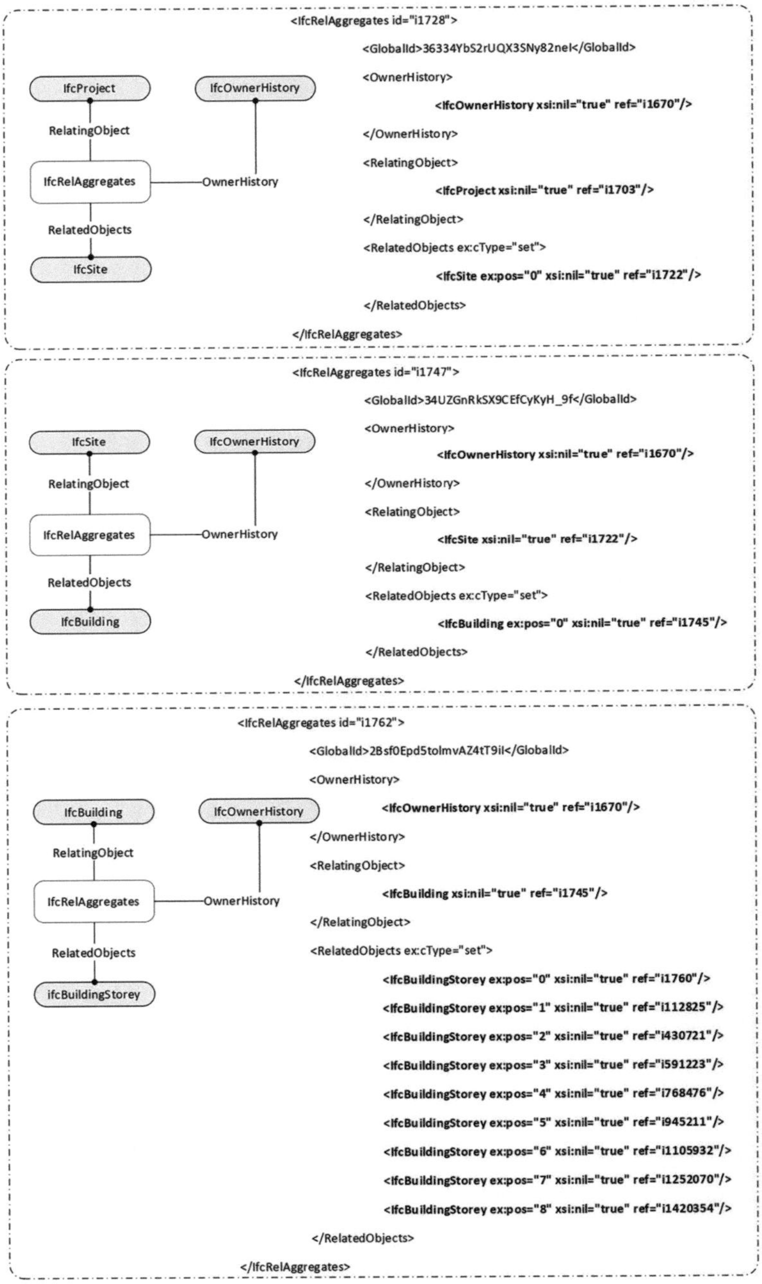

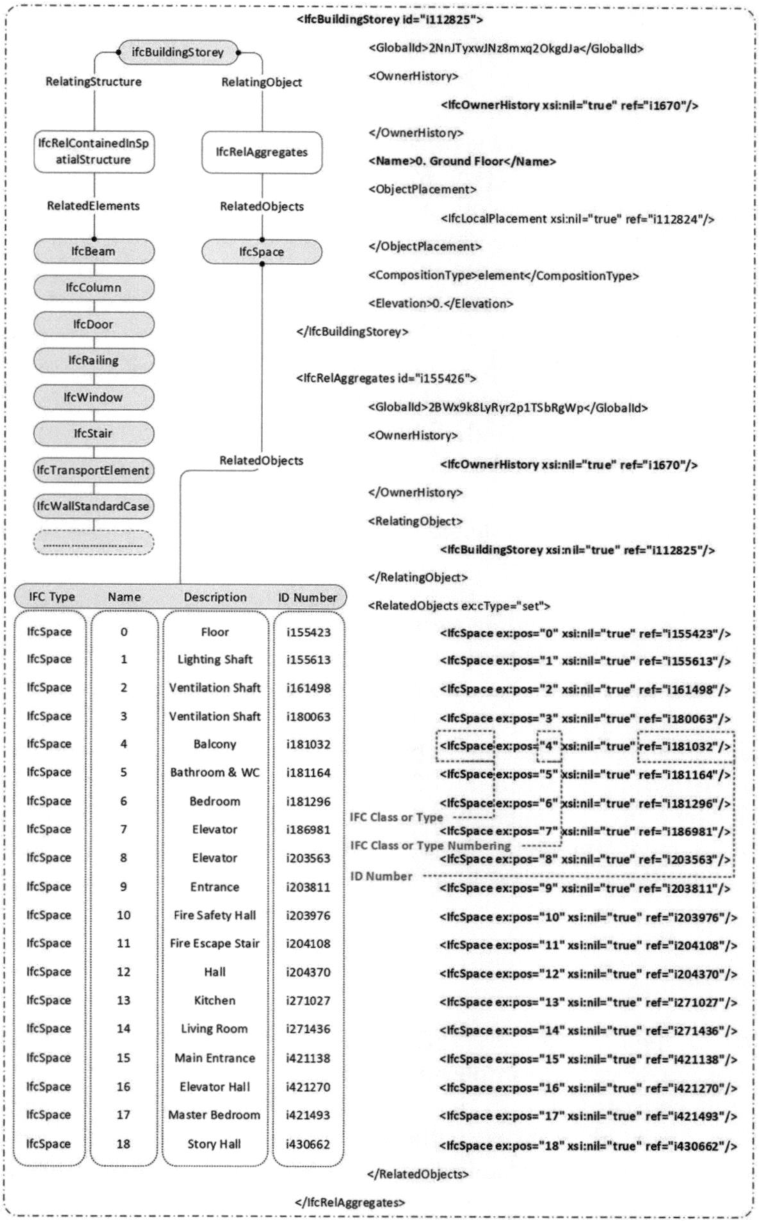

Figure 7.
The representation of the IFC class, the IFC class sorting, and ID number in IFCXML.

sorted with <RelatedObjects>. If there is more than one subclass, they are sorted by sequence number in the string like ex: pos = "0", ex: pos = "1", ex: pos = "2", etc. <RelatingObject> and <RelatedObjects> are always linked to <IfcOwnerHistory>. An <IfcRelAggregates> is used to sort and present all these relationships. An example of sorting subclasses in strings in IFC is shown in **Figure 7**.

An example of a hierarchical scheme of the IFC is shown in **Figure 7**. The abbreviations of the IFC data are given according to the IFC standard in **Figure 4**. The first equivalent of the sample housing project is IfcProject in IFC data. The logical relationships between IfcProject - IfcSite, IfcSite - IfcBuilding, and IfcBuilding - IfcBuildingStorey are provided IfcRelAggregates. IfcRelAggregates is where the data is sorted between the two IFC types. Since this data belongs to the building project, IfcRelAggregates is always linked to the IfcOwnerHistory representing the project owner. IfcRelAggregates makes the previous IFC type's logical relationship with RelatingObject and the next IFC type with RelatedObjects.

IfcProject contains necessary information about the building project. The IfcProject number must be one because the BIM model always represents a building project number. On the contrary, it is not possible to mention IFC for situations other than this. IfcSite contains information about the site, area, or plot on which the building project will be built. The building project can be built on more than one site. Therefore, the number of IFC types can be more than one except IfcProject. IfcBuilding involves the information of the building or buildings included in the building project. On the other hand, IfcBuildingStorey gives each floor information by sorting the number of floors in the building project. Finally, the information of building elements named IfcBuildingElement is listed.

The list of building elements included in the building project is put in order in two ways:

- IfcSpace represents a building element such as a room, space, or zone. IfcRelAggregates provide the logical relationship between IfcSpace and IfcBuildingStorey. All IfcSpace(s) are listed here via IfcRelAggregates.

- IfcBuildingElement represents all building elements except IfcSpace in IFC. The logical relationship between IfcBuildingElement and IfcBuildingStorey is provided by IfcRelContainedInSpatialStructure. Apart from IfcSpace, all other building elements such as IfcBeam, IfcColumn, IfcRailing, IfcSlab, IfcStair, IfcDoor, and IfcWindow are listed here via IfcRelContainedInSpatialStructure.

4. Conclusion

In this study, the BIM, IFC, and IFCXML representations of the building project data were explained over the sample housing project in the ACCC process. The sample housing project consisted of 9 floors, 8 flats, and 2 elevators. Flat plans on each floor were modeled differently from each other. It was completed with 179 rooms, 91 doors, 66 windows, 52 railings, 2 mechanical ventilation, 2 elevators, 8 stairs, 92 walls, 135 columns, 183 beams, and 95 slabs. The procedures to create the sample housing project with ArchiCAD software were described, respectively. The four attributes of the building elements of the sample housing project were defined with the IFC labels. The BIM model of the sample housing project was modeled. Finally, the IFCXML data file of the sample housing project was exported.

Before exporting to the IFCXML data file, all information regarding the building project must be modeled appropriately. When the BIM model is transformed into an IFC or IFCXML data file, building elements can be translated into the relevant IFC types. Transforming a building model into an IFC or IFCXML data file also provides the detailed information required for automated code compliance checking. This convenience allows the building regulation control, which the traditional method implements, to be controlled by the computer. It also brings the advantage that the building regulation control is checked over the IFC data file where the BIM model is exported, not on paper-based manually. This advantage is realized only in information technology systems and databases.

The IFC data standard is currently being developed, updated, and maintained by buildingSMART International [18]. buildingSMART develops open standards for BIM and openBIM. These open standards are related to data models, processes, and terms. Feature terms defined in the IFC specification depend on the terms defined in the buildingSMART data dictionary. buildingSMART collaborates with ISO to develop international, regional, and national standards. Today, when global standards are being accepted as national standards, both financial and technical support are provided to innovative studies in the field of automated code compliance checking by buildingSMART [20]. It is aimed to achieve the following objectives by buildingSMART:

- Developing and maintaining international standards for BIM such as IFC, IFCXML,
- Increasing the use of automated code compliance checking in building projects according to the building regulations,
- Providing low-cost data sharing and exchange, and
- Spreading and accelerating the use of buildingSMART technology throughout the building project life cycle in the construction industry.

Conflict of interest

The author declares no conflict of interest.

Notes/thanks/other declarations

The author received no specific funding for this study.

Author details

Murat Aydın
Graduate School of Natural and Applied Sciences, Ankara University, Ankara, Turkey

*Address all correspondence to: aydinmrt@ankara.edu.tr

References

[1] Aydın M, Yaman H. An overview of building information modelling (BIM) based building code compliance checking literature. Journal of Design. 2018;**14**(25):59-77

[2] Dimyadi J, Amor R. Automated building code compliance checking—Where is it at? In: Proc. 19th World Build. Congr. Constr. Soc. 5–9 May, 2013. pp. 172-185

[3] Martins JP, Monteiro A. LicA: A BIM based automated code-checking application for water distribution systems. Automation in Construction. 2013;**29**(23):12-23

[4] Aydın M. Building information modeling based automated building regulation compliance checking Asp.net web software. Intelligent Automation & Soft Computing. 2021;**28**(1):11-25

[5] Greenwood D, Lockley S, Malsane S, Matthews J. Automated compliance checking using building information models. In: Proc. Constr. Build. Real Estate Res. Conf. 2–3 September, 2010. pp. 363-371

[6] Fenves SJ, Garrett JH, Kiliccote H, Law KH, Reed KA. Computer representations of design standards and building codes: US perspective. International Journal of Constrution Information and Technology. 1995;**3**(1): 13-34

[7] Lee H, Lee J-K, Park S, Kim I. Translating building legislation into a computer-executable format for evaluating building permit requirements. Automation in Construction. 2016;**71**:49-61

[8] Ding L, Drogemuller R, Rosenman M, Marchant D, Gero J. Automating code checking for building designs—DesignCheck. In: Clients Driving Innovation: Moving Ideas into Practice. 2006. pp. 1-16

[9] Dimyadi J, Clifton C, Spearpoint M, Amor R. Regulatory knowledge encoding guidelines for automated compliance audit of building engineering design. In: Computing in Civil and Building Engineering (2014). 2014. pp. 536-543

[10] Shih S-Y, Sher W, Giggins H. Assessment of the building code of Australia to inform the development of BIM-enabled code checking system. In: Proc. 19th World Build. Congr. Constr. Soc; 5–9 May, 2013. pp. 1-12

[11] Aydın M, Yaman H. A literature review of automated code compliance checking concept. Journal of Design. 2020;**16**(29):79-97

[12] Aydın M, Yaman H. Domain knowledge representation languages and methods for building regulations. In: Communications in Computer and Information Science. 2020. pp. 101-121

[13] NBIMS-US. National BIM Standard —United States Version 3. Washington, DC: National Institute of Building Sciences; 2015

[14] Nawari NO, Alsaffar A. Understanding computable building codes. Civil Engineering Architect. 2015; **3**(6):163-171

[15] Nawari NO. Automating codes conformance. Journal of Architectural Engineering. 2012;**18**(4):315-323

[16] IAI. Release 1.0 IFC Model Architecture, International Alliance for Interoperability [Online]. 1997. Available from: http://www.

buildingsmart-tech.org/ifc/ [Accessed: January 1, 2018]

[17] ISO 10303-11. Industrial Automation Systems and Integration—Product Data Representation and Exchange Description Methods: The EXPRESS Language Reference Manual. ISO Central Secretariat; 1997

[18] BuildingSMART. buildingSMART International [Online]. 2020. Available from: https://www.buildingsmart.org [Accessed: January 1, 2020]

[19] Graphisoft. ArchiCAD 21 [Online]. 2018. Available from: https://www.graphisoft.com/archicad/ [Accessed: December 1, 2018]

[20] buildingSmart Türkiye. buildingSmart Türkiye [Online]. 2019. Available from: http://www.buildingsmarttturkiye.org/index.php/hakkinda/buildingsmart-turkiye [Accessed: January 1, 2020]